Power PRACTICE™

MW01707164

Word Games

4–5

Written by
Linda Schwartz

Editor: Carla Hamaguchi
Illustrator: Jane Yamada
Designer/Production: Moonhee Pak/Terri Lamadrid
Cover Designer: Barbara Peterson
Art Director: Tom Cochrane
Project Director: Carolea Williams

Table of Contents

What Would You Do?

Ladder Links

Analogies

Fun with Words

Introduction

Each book in the *Power Practice*™ series contains dozens of ready-to-use activity pages to provide students with skill practice. Use the fun activities to supplement and enhance what you are already teaching in your classroom. Give an activity page to students as independent class work, or send the pages home as homework to reinforce skills taught in class. An answer key is provided for quick reference.

Word Games 4–5 is filled with challenging puzzles and word games to help students improve their vocabulary and spelling as well as sharpen their creative thinking skills. The book is divided into nine main sections:

- **Make-a-Word Puzzles**
  Students go from letter to letter in adjacent circles or squares to see how many words they can make.
- **Word Detectives**
  Students start with holiday or special occasion words such as *Happy Halloween*. They use the letters of these words to write as many words as they can.
- **Word Search Puzzles**
  These word search puzzles are correlated to science and social studies topics. Students find and circle the 20 hidden words going forward, backward, up, down, and diagonally. Each puzzle also features an extended learning activity.
- **Category Game**
  Students list words beginning with specific letters of the alphabet for five categories. They are encouraged to use reference materials to find names of unusual mountains, rivers, authors, etc. Students can exchange game sheets and easily check answers.
- **Crossword Puzzles**
  These crossword puzzles are correlated to parts of speech, synonyms, antonyms, homophones, plurals, abbreviations, and holidays.
- **What Would You Do?**
  These fun puzzle sheets encourage students to use the dictionary to look up unusual words.
- **Ladder Links**
  Students change one word at the top of a ladder into another word at the bottom of the ladder by changing only one letter at a time. Definition clues help them solve the puzzles.
- **Analogies**
  Students complete an analogy by deciding on the best word choice.
- **Fun with Words**
  This is a collection of puzzles, including palindromes, picture word puzzles, word scrambles, and more!

On many of the pages, a Just for Fun activity is provided for extended learning. Use the ready-to-go activities to "recharge" skill review and give students the power to succeed!

Name ______________________________ Date ______________

Ferris Wheel Fun

Spelling and Vocabulary

How many words can you find in the Ferris wheel? Write the words you find on a separate sheet of paper. Hint: There are more than **35** words!

Rules

- The words you find must have three or more letters.
- Start on any circle and move to an adjacent circle connected by a line to spell a word. You may not skip a circle.

  Examples: *Tin* is allowed because each circle is next to the other and is connected by a line.
  Pine is not allowed because the letter *n* is not connected to the letter *e*.
- You may not use the same letter twice in a row, but you can go back and use a letter again in the same word.
- Plurals (e.g., *tiers*) are allowed.
- No proper nouns are allowed.

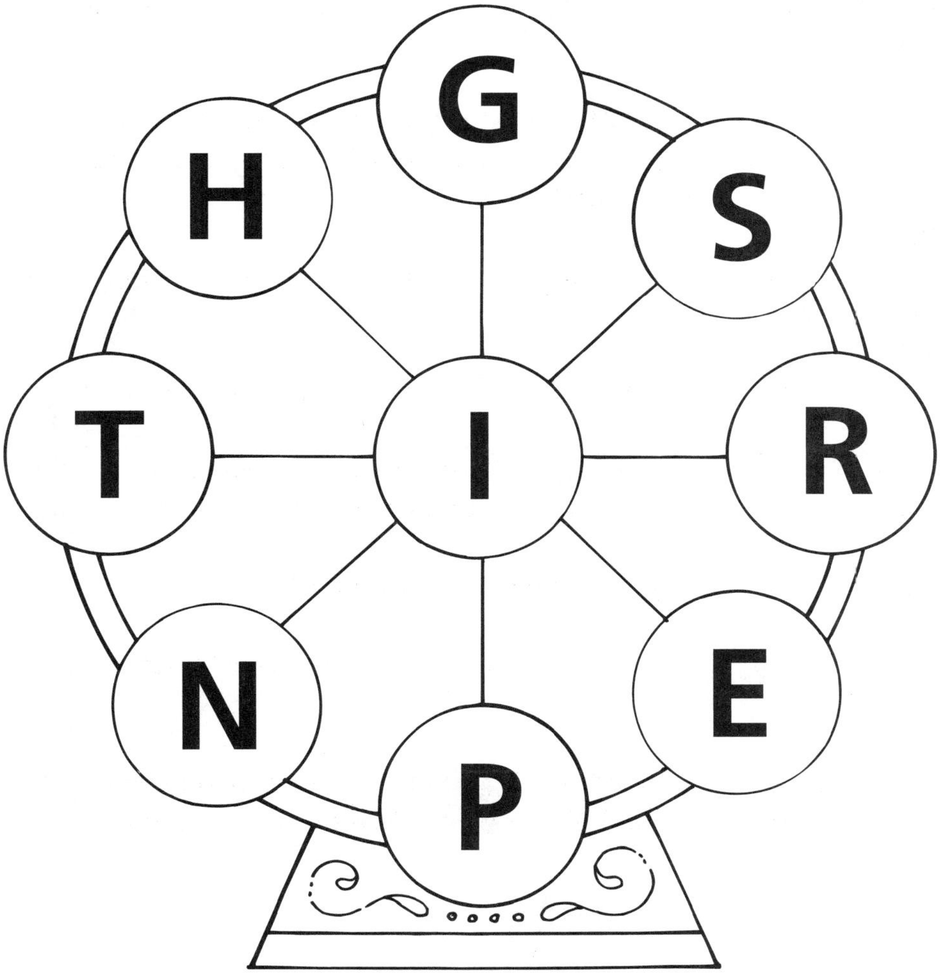

Name ______________________________ Date ______________

Circle Words

SPELLING AND VOCABULARY

How many words can you find in the connected circles? Write the words you find on a separate sheet of paper. Hint: There are more than **50** words!

Rules

- The words you find must have three or more letters.
- Start on any circle and move to an adjacent circle connected by a line to spell a word. You may not skip a circle.

  Examples: *How* is allowed because each circle is next to each other.
  Hot is not allowed because the letter *t* is not connected to the letter *o*.
- You may not use the same letter twice in a row (e.g., *arrest*), but you can go back and use a letter again in the same word (e.g., *rear*).
- Plurals (e.g., *rests*) are allowed.
- No proper nouns (e.g., *Stewart*) are allowed.

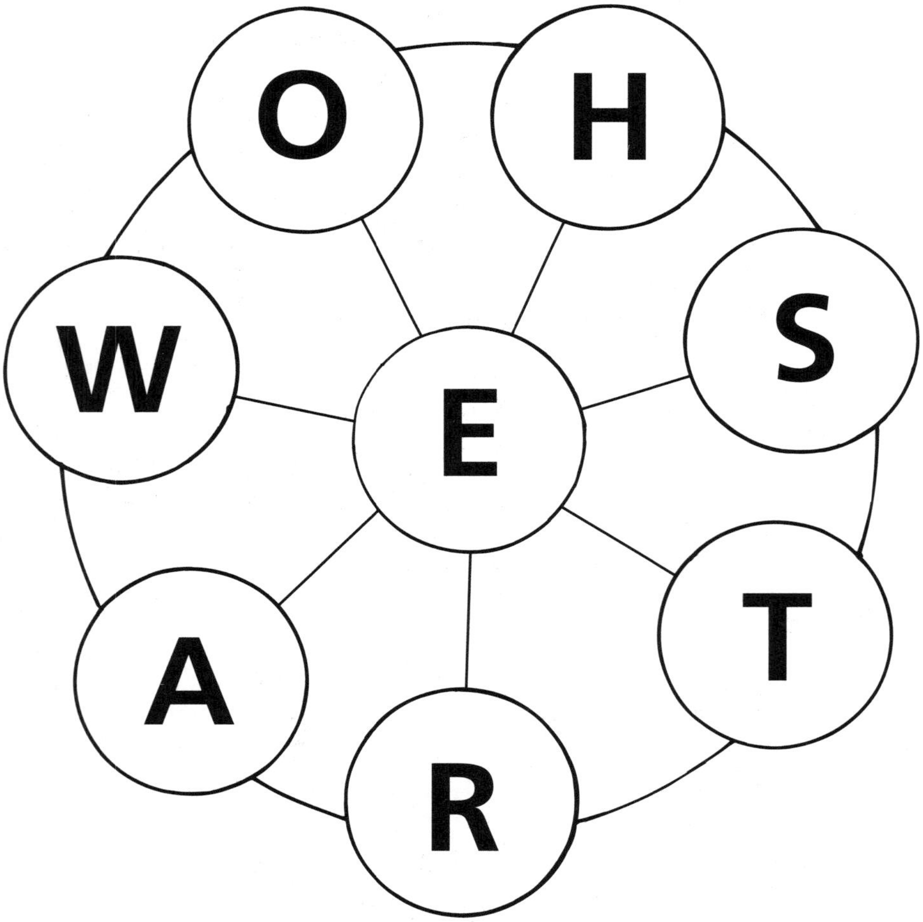

Name ______________________ Date ____________

Pepperoni Pizza

SPELLING AND VOCABULARY

How many words can you find in the pizza? Write the words you find on a separate sheet of paper. Hint: There are more than **35** words and at least one eight-letter word!

Rules

- The words you find must have three or more letters.
- Start on any circle (pepperoni) and move to an adjacent circle connected by a line to spell a word. You may not skip a circle.

  Examples: *Sip* is allowed because each adjacent pepperoni is connected by a line.
  Tap is not allowed because the letter *a* is not connected to the letter *p*.
- You may not use the same letter twice in a row (e.g., *tarry*), but you can go back and use a letter again in the same word (e.g., *pep*).
- Plurals (e.g., *rats*) are allowed.
- No proper nouns are allowed.

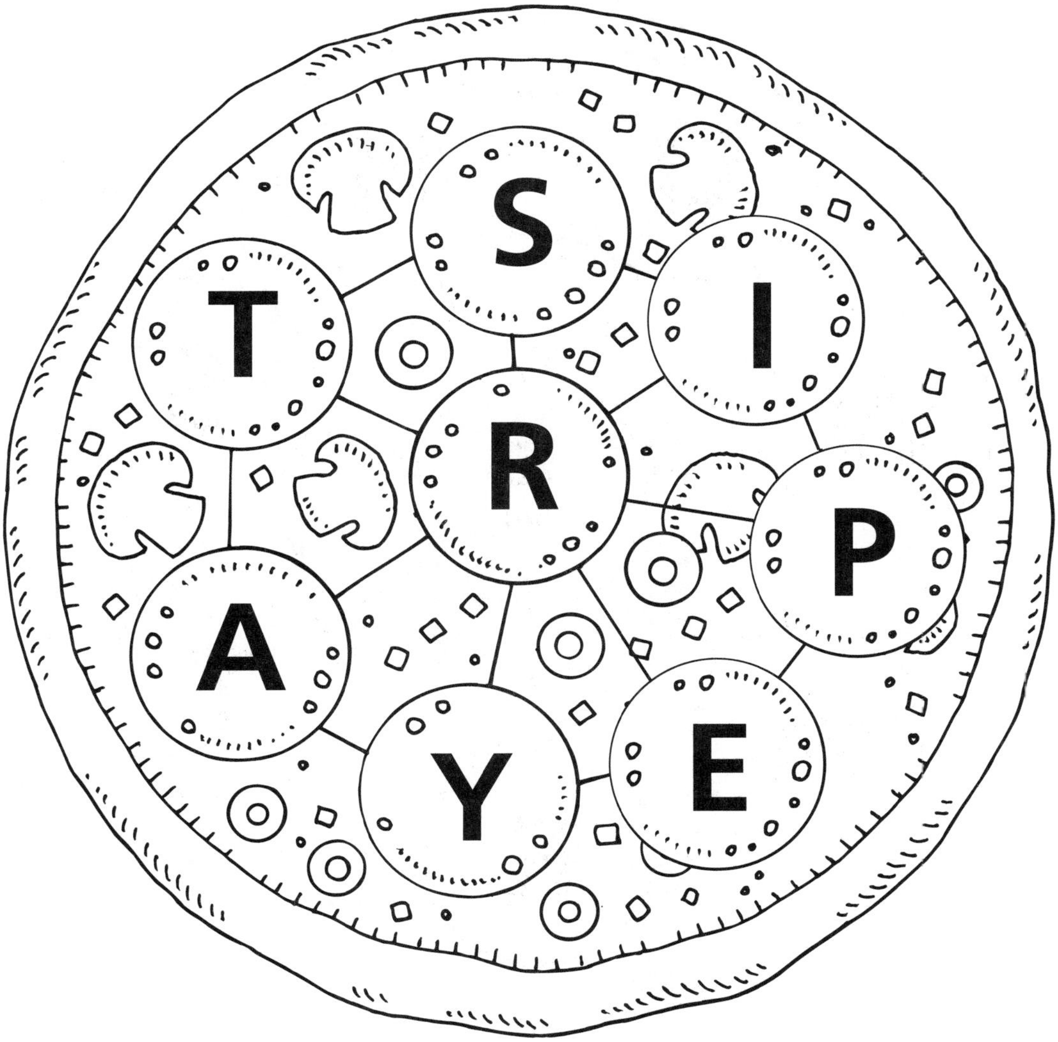

Name ______________________ Date ____________

Pick a Petal

SPELLING AND VOCABULARY

How many words can you find in the flower? Write the words you find on a separate sheet of paper. Hint: There are more than **35** words!

Rules

- The words you find must have three or more letters.
- Start on any part of the flower (center or petals) and move to an adjacent part to spell a word. You may not skip a petal.

  Examples: *Bet* is allowed because each petal is touching another petal or the letter in the center.

  Pit is not allowed because the petals with the letters *i* and *t* are not touching each other.
- You may not use the same letter twice in a row (e.g., *bell*), but you can go back and use a letter again in the same word (e.g., *pep*).
- Plurals (e.g., *beds*) are allowed.
- No proper nouns are allowed.

Name ______________________________ Date ______________

More Pick a Petal

SPELLING AND VOCABULARY

How many words can you find in the flower? Write the words you find on a separate sheet of paper. Hint: There are more than **55** words!

Rules

- The words you find must have three or more letters.
- Start on any part of the flower (center or petals) and move to an adjacent part to spell a word. You may not skip a petal.
  Examples: *Near* is allowed because each petal is touching another petal or the letter in the center.
  Name is not allowed because the petals with the letters *m* and *e* are not touching each other.
- You may not use the same letter twice in a row (e.g., *teen*), but you can go back and use a letter again in the same word (e.g., *eaten*).
- Plurals (e.g., *rats*) are allowed.
- No proper nouns (e.g., *Pam*) are allowed.

Name ______________________________ Date ______________

Word Quilt

Spelling and Vocabulary

How many words can you find in the quilt? Write the words you find on a separate sheet of paper. Hint: There are more than **90** words!

Rules

- The words you find must have three or more letters.
- Start on any square of the quilt and move one square in any direction to spell a word. You may not skip a square.

  Examples: *Rat* is allowed because each square is touching another square.
  Ran is not allowed because the squares with the letters *a* and *n* are not touching each other.
- You may not use the same letter twice in a row (e.g., *tarry*), but you can go back and use a letter again in the same word (e.g., *tartar*).
- Plurals (e.g., *beds*) are allowed.
- No proper nouns (e.g., *Ted*) are allowed.

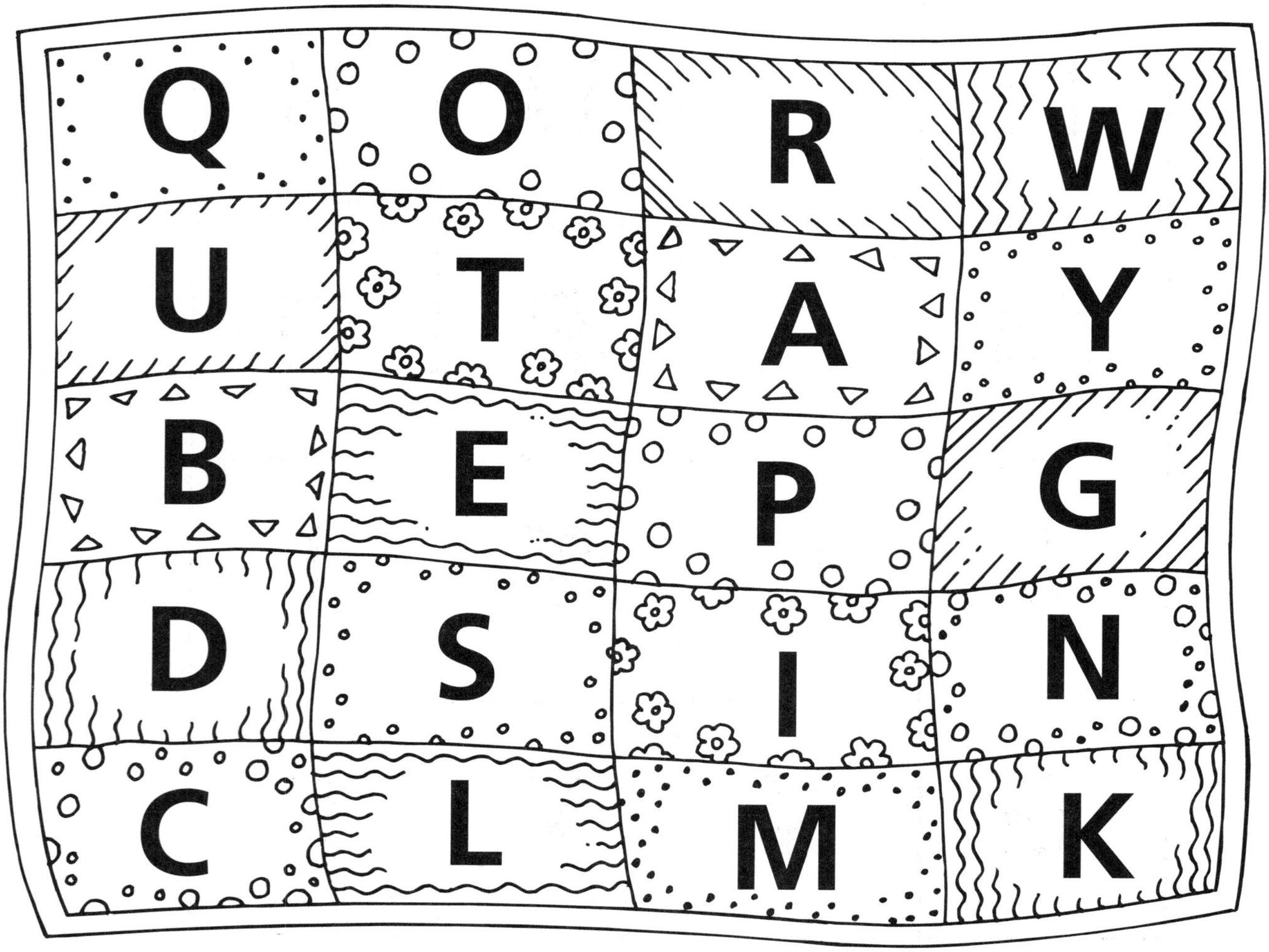

Name ______________________________ Date ______________

Magic Square #1

Spelling and Vocabulary

How many words can you find in the magic square? Write the words you find on a separate sheet of paper. Hint: There are more than **75** words!

Rules

- The words you find must have three or more letters.
- Start on any square and move up, down, left, right, or diagonally from square to square to spell a word. You may not skip a square.

  Examples: *Care* is allowed because each square is touching another square.
  Learn is not allowed because the squares with the letters *r* and *n* are not touching each other.
- You may not use the same letter twice in a row (e.g., *boot*), but you can go back and use a letter again in the same word (e.g., *rarer*).
- No proper nouns (e.g., *Burt*) are allowed.

H	C	U	B
M	A	R	O
L	E	T	I
G	N	Y	P

Name ______________________ Date ______________

Magic Square #2

SPELLING AND VOCABULARY

How many words can you find in the magic square? Write the words you find on a separate sheet of paper. Hint: There are more than **90** words!

Rules

- The words you find must have three or more letters.
- Start on any square and move up, down, left, right, or diagonally from square to square to spell a word. You may not skip a square.

  Examples: *Ring* is allowed because each square is touching another square.
  Nice is not allowed because the squares with the letters *i* and *c* are not touching each other.
- You may not use the same letter twice in a row (e.g., *good*), but you can go back and use a letter again in the same word (e.g., *nine*).
- Plurals (e.g., *dogs*) are allowed.
- No proper nouns (e.g., *Diane*) are allowed.

R	D	O	T
P	I	G	S
Y	A	N	U
M	C	E	B

Name ______________________________ Date ______________

Magic Square #3

Spelling and Vocabulary

How many words can you find in the magic square? Write the words you find on a separate sheet of paper. Hint: There are more than **85** words!

Rules

- The words you find must have three or more letters.
- Start on any square and move up, down, left, right, or diagonally from square to square to spell a word. You may not skip a square.

  Examples: *For* is allowed because each square is touching another square.
  Rot is not allowed because the squares with the letters *o* and *t* are not touching each other.
- You may not use the same letter twice in a row (e.g., *door*), but you can go back and use a letter again in the same word (e.g., *lava*).
- Plurals (e.g., *words*) are allowed.
- No proper nouns are allowed.

K	F	D	S	U
N	O	R	T	H
E	W	L	A	M
C	Y	I	V	B

Name ______________________________ Date ______________

Magic Square #4

Spelling and Vocabulary

How many words can you find in the magic square? Write the words you find on a separate sheet of paper. Hint: There are more than **100** words!

Rules

- The words you find must have three or more letters.
- Start on any square and move up, down, left, right, or diagonally from square to square to spell a word. You may not skip a square.
  Examples: *Lot* is allowed because each square is touching another square.
  For is not allowed because the squares with the letters *o* and *r* are not touching each other.
- You may not use the same letter twice in a row (e.g., *foot*), but you can go back and use a letter again in the same word (e.g., *tot*).
- Plurals (e.g., *taps*) are allowed.
- No proper nouns are allowed.

G	M	L	F
N	I	T	O
E	S	A	H
R	V	P	W

Name ______________________ Date ____________

Valentine's Day

SPELLING AND VOCABULARY

How many words can you make using the letters in *Valentine's Day*? Write the words on the lines below. Hint: There are more than **150** words!

Rules
- The words you make must have three or more letters.
- Your words can use letters in any order from *Valentine's* and *Day*.
- You may only use each letter as many times as it appears in the two words.
  Example: You may not list the word *tell* because there is only one letter *l*.
- Plurals (e.g., *days*) are allowed.
- No proper nouns (e.g., *Ted*) are allowed.

Scoring

Three-letter word	= 1 point
Four-letter word	= 2 points
Five-letter word	= 3 points
Six-letter word	= 4 points
Word with seven or more letters	= 5 points

Number of words I found: ____________ Points: ____________

Name ______________________ Date ____________

Abraham Lincoln

SPELLING AND VOCABULARY

How many words can you make using the letters in *Abraham Lincoln*? Write the words on the lines below. Hint: There are more than **50** words!

Rules

- The words you make must have three or more letters.
- Your words can use letters in any order from *Abraham* and *Lincoln*.
- You may only use each letter as many times as it appears in the two words.

  Example: You may not list the word *loom* because there is only one letter *o*.
- No proper nouns (e.g., *Lincoln*) are allowed.

<u>Scoring</u>

Three-letter word	= 1 point
Four-letter word	= 2 points
Five-letter word	= 3 points
Six-letter word	= 4 points
Word with seven or more letters	= 5 points

Number of words I found: __________ Points: __________

Name ______________________________ Date ______________

George Washington

Spelling and Vocabulary

How many words can you make using the letters in *George Washington*? Write the words on the lines below. Hint: There are more than **130** words!

Rules

- The words you make must have three or more letters.
- Your words can use letters in any order from *George* and *Washington*.
- You may only use each letter as many times as it appears in the two words.
  Example: You may not list the word *washes* because there is only one letter *s*.
- No proper nouns (e.g., *Washington*) are allowed.

Scoring

Three-letter word	= 1 point
Four-letter word	= 2 points
Five-letter word	= 3 points
Six-letter word	= 4 points
Word with seven or more letters	= 5 points

Number of words I found: ______________ Points: ______________

Name ______________________ Date ____________

Mother's Day

Spelling and Vocabulary

How many words can you make using the letters in *Mother's Day*? Write the words on the lines below. Hint: There are more than **105** words!

Rules

- The words you make must have three or more letters.
- Your words can use letters in any order from *Mother's* and *Day*.
- You may only use each letter as many times as it appears in the two words.
  Example: You may not list the word *mom* because there is only one letter *m*.
- Plurals (e.g., *days*) are allowed.
- No proper nouns (e.g., *Tom*) are allowed.

Scoring

Three-letter word	= 1 point
Four-letter word	= 2 points
Five-letter word	= 3 points
Six-letter word	= 4 points
Word with seven or more letters	= 5 points

Number of words I found: ____________ Points: ____________

Name ______________________ Date ____________

Father's Day

Spelling and Vocabulary

How many words can you make using the letters in *Father's Day*? Write the words on the lines below. Hint: There are more than **100** words!

Rules

- The words you make must have three or more letters.
- Your words can use letters in any order from *Father's* and *Day*.
- You may only use each letter as many times as it appears in the two words.
  Example: You may not list the word *that* because there is only one letter *t*.
- Plurals (e.g., *days*) are allowed.
- No proper nouns (e.g., *Thad*) are allowed.

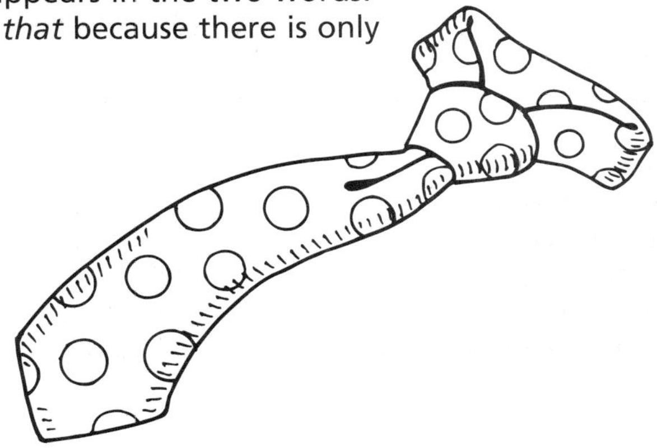

Scoring

Three-letter word	= 1 point
Four-letter word	= 2 points
Five-letter word	= 3 points
Six-letter word	= 4 points
Word with seven or more letters	= 5 points

Number of words I found: ____________ Points: ____________

Name ______________________________ Date ______________

Summer Vacation

SPELLING AND VOCABULARY

How many words can you make using the letters in *summer vacation*? Write the words on the lines below. Hint: There are more than **100** words!

Rules

- The words you make must have three or more letters.
- Your words can use letters in any order from *summer* and *vacation*.
- You may only use each letter as many times as it appears in the two words.
  Example: You may not list the word *rare* because there is only one letter *r*.
- Plurals (e.g., *cats*) are allowed.
- No proper nouns (e.g., *Vic*) are allowed.

Scoring

Three-letter word	= 1 point
Four-letter word	= 2 points
Five-letter word	= 3 points
Six-letter word	= 4 points
Word with seven or more letters	= 5 points

Number of words I found: ____________ Points: ______________

Name ______________________________ Date ______________

Independence Day

SPELLING AND VOCABULARY

How many words can you make using the letters in *Independence Day*? Write the words on the lines below. Hint: There are more than **65** words!

Rules

- The words you make must have three or more letters.
- Your words can use letters in any order from *Independence* and *Day*.
- You may only use each letter as many times as it appears in the two words.
  Example: You may not list the word *pappy* because there is only one letter *p*.
  The word *added* is allowed because there are three *d*s in *In**d**epen**d**ence **D**ay*.
- No proper nouns (e.g., *Dan*) are allowed.

Scoring

Three-letter word	= 1 point
Four-letter word	= 2 points
Five-letter word	= 3 points
Six-letter word	= 4 points
Word with seven or more letters	= 5 points

Number of words I found: ______________ Points: ______________

Name ______________________________ Date ______________

Happy Halloween

Spelling and Vocabulary

How many words can you make using the letters in *Happy Halloween*? Write the words on the lines below. Hint: There are more than **90** words!

Rules

- The words you make must have three or more letters.
- Your words can use letters in any order from *Happy* and *Halloween*.
- You may only use each letter as many times as it appears in the two words.
  Example: You may not list the word *nanny* because there is only one letter *n*.
- No proper nouns (e.g., *Hal*) are allowed.

Scoring

Three-letter word	= 1 point
Four-letter word	= 2 points
Five-letter word	= 3 points
Six-letter word	= 4 points
Word with seven or more letters	= 5 points

____________	____________	____________
____________	____________	____________
____________	____________	____________
____________	____________	____________
____________	____________	____________
____________	____________	____________
____________	____________	____________
____________	____________	____________
____________	____________	____________

Number of words I found: ____________ Points: ____________

Name ______________________________ Date ______________

Festival of Lights

SPELLING AND VOCABULARY

How many words can you make using the letters in *Festival of Lights*? Write the words on the lines below. Hint: There are more than **200** words!

Rules

- The words you make must have three or more letters.
- Your words can use letters in any order from *Festival*, *of*, and *Lights*.
- You may only use each letter as many times as it appears in the three words.
  Example: You may not list the word *gag* because there is only one letter *g*.
- No proper nouns (e.g., *Seth*) are allowed.

Scoring

Three-letter word	= 1 point
Four-letter word	= 2 points
Five-letter word	= 3 points
Six-letter word	= 4 points
Word with seven or more letters	= 5 points

Number of words I found: ____________ Points: ______________

Name ______________________________ Date ______________

Merry Christmas

SPELLING AND VOCABULARY

How many words can you make using the letters in *Merry Christmas*? Write the words on the lines below. Hint: There are more than **125** words!

Rules

- The words you make must have three or more letters.
- Your words can use letters in any order from *Merry* and *Christmas*.
- You may only use each letter as many times as it appears in the two words.
  Example: You may not list the word *matter* because there is only one letter *t*.
- Plurals (e.g., *ears*) are allowed.
- No proper nouns (e.g., *Christmas*) are allowed.

Scoring

Three-letter word	= 1 point
Four-letter word	= 2 points
Five-letter word	= 3 points
Six-letter word	= 4 points
Word with seven or more letters	= 5 points

Number of words I found: ______________ Points: ________________

Name ______________________________ Date ______________

Bone Search

SPELLING AND SCIENCE VOCABULARY

Find and circle the names of the bones hidden in the puzzle. Words can be found going forward, backward, up, down, and diagonally.

clavicle	coccyx	femur	fibula	humerus
ilium	lumbar	vertebrae	mandible	
maxilla	metacarpus	metatarsus	patella	
pelvis	radius	sacrum	scapula	
sternum	tibia	ulna	tarsus	

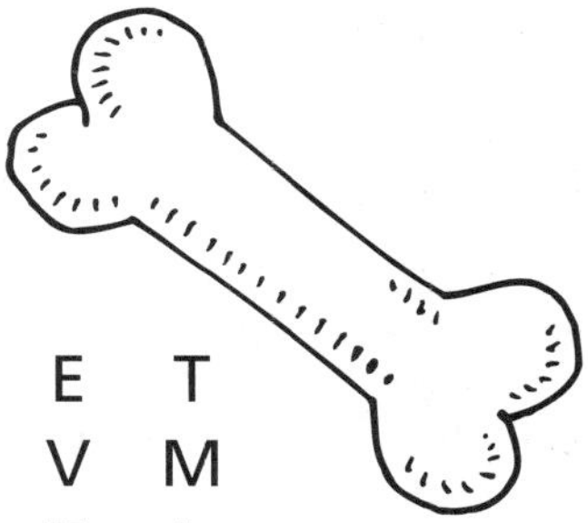

E	C	E	L	E	V	B	S	S	R	T	L	E	T
A	E	L	E	E	S	C	O	C	C	Y	X	V	M
R	P	F	M	S	U	I	D	A	R	L	E	T	L
B	A	R	U	E	S	E	S	P	S	A	I	R	E
E	T	U	S	U	R	E	M	U	H	B	L	E	A
T	E	U	L	I	A	N	P	L	I	E	I	I	N
R	L	B	T	M	T	R	U	A	I	L	E	P	U
E	L	C	I	V	A	L	C	L	N	E	E	I	U
V	A	S	N	C	T	N	I	U	U	L	P	A	R
R	U	E	A	T	E	U	D	B	V	A	U	T	P
A	U	T	R	C	M	A	X	I	L	L	A	A	U
B	E	M	I	U	R	U	S	F	B	R	S	A	C
M	R	L	E	E	T	U	T	R	S	L	S	B	N
U	A	A	M	F	R	U	M	U	N	R	E	T	S
L	B	D	B	N	M	Y	S	R	R	C	R	B	M

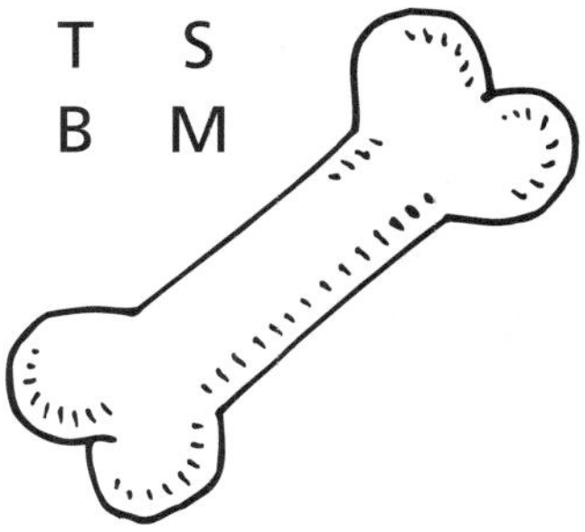

Just for Fun

Select eight bones from the puzzle. Tell where each bone is located in the human body.

Name ______________________________ Date ______________

Color Search

SPELLING AND SCIENCE VOCABULARY

Find and circle the names of the colors hidden in the puzzle. Words can be found going forward, backward, up, down, and diagonally.

amber	cobalt	emerald	lavender	taupe
auburn	coral	hazel	magenta	teal
beige	crimson	indigo	ruby	turquoise
chartreuse	ebony	khaki	scarlet	vermilion

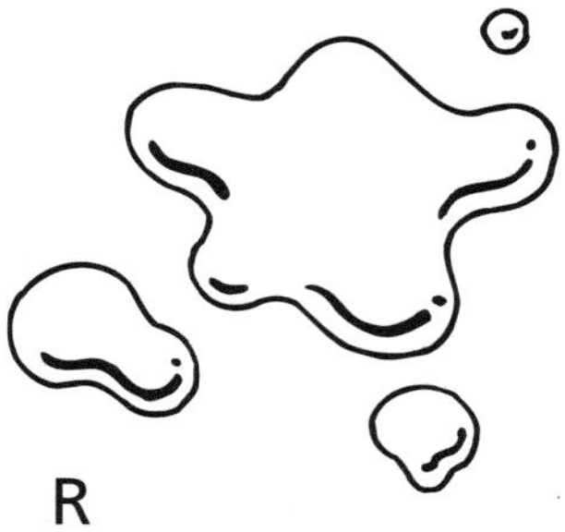

A	E	P	U	A	T	L	E	E	C	I	R
T	D	B	B	I	E	H	Y	E	E	R	E
N	U	L	O	Z	U	B	Y	S	O	I	E
E	T	R	A	N	U	V	M	C	I	K	N
G	A	H	Q	R	Y	E	R	A	L	A	E
A	A	E	S	U	E	R	T	R	A	H	C
M	N	T	O	B	O	M	E	L	V	K	E
B	Y	R	L	U	I	I	E	E	E	A	S
E	N	A	E	A	K	L	S	T	N	L	R
R	E	G	I	E	B	I	O	E	D	M	A
T	I	N	D	I	G	O	I	S	E	R	E
C	R	I	M	S	O	N	C	O	R	A	L

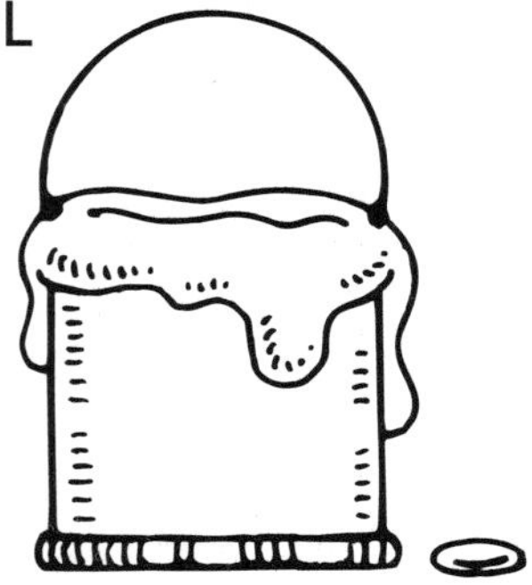

Just for Fun
Make a graph with the following headings: red, yellow, blue, green, black, orange, purple, and brown. Then list each color in the puzzle under the correct heading. Use a dictionary if you need help.

Name ______________________________ Date ______________

Dog Search

SPELLING AND SCIENCE VOCABULARY

Find and circle the breeds of dogs hidden in the puzzle. Words can be found going forward, backward, up, down, and diagonally.

Australian shepherd	bulldog	dalmatian	Irish setter
basset hound	Chihuahua	German shepherd	Labrador retriever
beagle	cocker spaniel	golden retriever	poodle
bloodhound	collie	Great Dane	terrier
boxer	dachshund	greyhound	whippet

D	E	I	D	N	U	O	H	Y	E	R	G	T	B	R	R	D
R	O	R	R	C	O	L	L	I	E	L	H	A	E	C	N	O
E	B	C	E	A	T	B	B	T	N	S	G	D	P	P	L	O
H	E	P	H	S	S	E	T	A	D	E	N	A	U	E	R	B
P	R	I	P	R	L	E	P	S	R	U	B	R	E	E	V	H
E	H	A	E	L	S	I	D	P	H	O	B	O	V	B	A	E
H	R	K	H	H	A	A	D	S	I	S	S	E	X	E	T	A
S	D	D	S	E	A	U	H	A	U	H	I	H	C	E	E	E
N	G	I	N	U	R	C	A	I	N	R	W	T	E	L	R	E
A	R	O	A	B	A	S	S	E	T	H	O	U	N	D	R	N
I	E	S	M	D	C	R	L	E	E	R	S	A	E	T	I	A
L	A	B	R	A	D	O	R	R	E	T	R	I	E	V	E	R
A	T	U	E	E	D	N	U	O	H	D	O	O	L	B	R	I
R	D	L	G	L	E	I	N	A	P	S	R	E	K	C	O	C
T	A	L	R	D	A	L	M	A	T	I	A	N	T	R	R	A
S	N	D	L	O	T	B	G	T	S	H	G	H	E	N	V	N
U	E	O	E	O	N	I	H	G	A	N	C	S	N	H	R	N
A	G	G	S	P	L	I	N	E	L	L	H	T	H	E	G	R

Just for Fun
Create a word search (with an answer key) that includes the names of 20 additional breeds of dogs. Give your puzzle to a classmate to complete.

Name ______________________________ Date ______________

Flower Search

SPELLING AND SCIENCE VOCABULARY

Find and circle the names of the flowers hidden in the puzzle. Words can be found going forward, backward, up, down, and diagonally.

azalea	daffodil	iris	pansy	rose
begonia	dandelion	lilac	petunia	sunflower
buttercup	daisy	marigold	poppy	tulip
carnation	gardenia	orchid	rhododendron	violet

E	I	T	O	U	L	A	O	I	O	R	A
L	P	A	N	S	Y	N	V	E	O	T	D
U	O	L	O	N	P	I	L	U	T	N	O
O	P	L	I	D	O	F	F	A	D	C	I
N	P	B	L	L	E	E	Z	O	A	R	D
O	Y	U	E	A	A	A	R	E	I	F	L
I	N	T	D	O	L	C	I	S	S	R	O
T	Y	T	N	E	H	F	L	O	Y	T	G
A	O	E	A	I	N	E	D	R	A	G	I
N	O	R	D	N	E	D	O	D	O	H	R
R	N	C	I	U	B	E	G	O	N	I	A
A	S	U	N	F	L	O	W	E	R	B	M
C	D	P	E	T	U	N	I	A	Y	O	O

Just for Fun
Draw pictures of six of these flowers, and label each one. Write two fascinating facts about each flower.

Name ______________________________ Date ________________

Machine Search

Spelling and Science Vocabulary

Find and circle the names of the machines hidden in the puzzle. Words can be found going forward, backward, up, down, and diagonally.

battering ram	engine	pump	waterwheel
block and tackle	inclined plane	screw	wedge
crane	lever	steam shovel	wheel and axle
derrick	machine tool	treadmill	windlass
electric generator	pulley	turbine	windmill

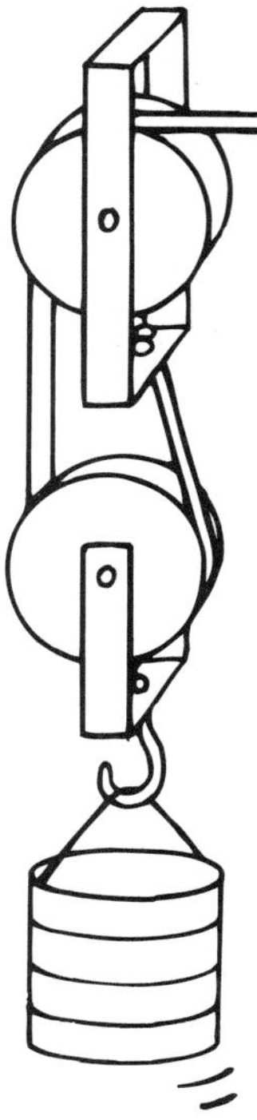

R	Y	W	E	D	G	E	A	N	T	A	D	U	N	H	L
O	H	I	N	T	E	L	E	E	H	W	R	E	T	A	W
T	C	L	R	A	M	X	N	M	N	A	T	I	L	T	I
A	T	U	R	L	A	A	M	A	U	G	N	I	U	B	N
R	R	L	N	O	R	D	A	C	K	C	I	R	R	E	D
E	I	S	L	C	L	N	R	H	L	B	B	N	E	A	L
N	I	L	L	E	L	A	G	I	L	I	K	S	E	R	A
E	P	E	L	H	I	L	N	N	N	G	E	T	S	E	S
G	E	M	P	L	M	E	I	E	S	C	R	E	W	T	S
C	E	I	U	E	D	E	R	T	E	B	M	A	T	R	A
I	O	E	L	P	A	H	E	O	C	R	E	M	I	D	O
R	O	O	L	R	E	W	T	O	E	E	W	S	M	P	E
T	T	A	E	I	R	K	T	L	V	V	E	H	A	E	E
C	N	V	Y	E	T	E	A	U	D	A	R	O	P	E	R
E	E	I	I	E	I	A	B	T	D	C	E	V	A	N	R
L	L	I	M	D	N	I	W	E	I	E	L	E	H	B	N
E	L	K	C	A	T	D	N	A	K	C	O	L	B	K	D

Just for Fun
Select the names of six machines from the puzzle. Draw and label diagrams of these machines to show how they work.

Name ______________________________ Date ______________

Mammal Search

SPELLING AND SCIENCE VOCABULARY

Find and circle the names of the mammals hidden in the puzzle. Words can be found going forward, backward, up, down, and diagonally.

anteater	chipmunk	moose	panda	rhinoceros
baboon	coyote	ocelot	porcupine	skunk
camel	gorilla	opossum	rabbit	walrus
cheetah	hippopotamus	otter	raccoon	wolf

S	T	C	A	M	E	L	N	B	I	T	O
K	U	O	C	H	E	E	T	A	H	A	S
U	C	C	O	H	T	N	S	B	T	A	O
N	A	E	A	O	I	I	W	O	E	R	R
K	P	L	Y	A	A	P	E	O	O	S	E
A	P	O	C	N	L	U	M	N	L	M	C
H	C	T	M	T	L	C	U	U	S	F	O
C	E	M	R	E	I	R	S	O	N	T	N
T	I	B	B	A	R	O	S	O	N	K	I
S	U	M	A	T	O	P	O	P	P	I	H
L	O	A	W	E	G	C	P	C	R	R	R
O	T	T	E	R	N	O	O	C	C	A	R
W	A	L	R	U	S	A	D	N	A	P	A

Just for Fun
Write a short story in which you combine the characteristics of two of these mammals. For example, you might write a humorous story about a mammal with the tusks of a walrus and the quills of a porcupine.

Name ______________________________ Date ______________

Space Search

Spelling and Science Vocabulary

Find and circle the space words hidden in the puzzle. Words can be found going forward, backward, up, down, and diagonally.

apogee	black hole	dwarf	nebula	sunspots
asteroid	comet	eclipse	penumbra	supernova
astronomy	constellation	galaxy	planet	umbra
aurora	corona	meteorite	satellite	zenith

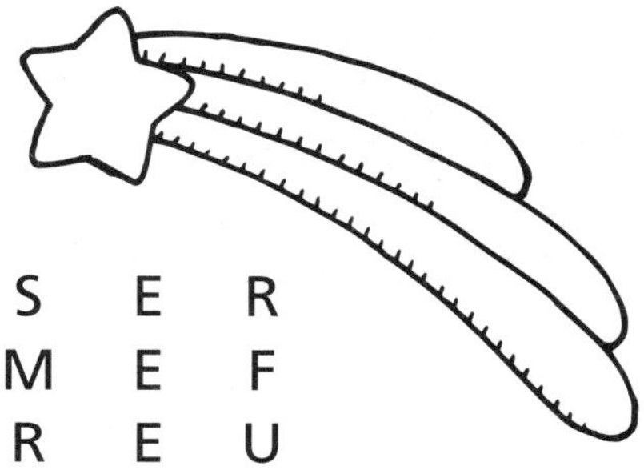

G	A	L	A	X	Y	P	P	I	L	S	E	R
M	U	Z	E	S	P	I	L	C	E	M	E	F
A	Y	E	T	O	U	R	N	T	E	R	E	U
E	M	N	A	O	S	P	I	T	H	L	G	R
N	O	I	T	A	L	L	E	T	S	N	O	C
D	N	T	T	A	L	O	L	R	U	D	P	O
S	O	H	N	E	R	B	O	U	N	I	A	R
A	R	E	T	I	M	D	H	U	S	O	S	O
Y	T	A	T	N	W	O	K	M	P	R	V	N
M	S	E	I	A	W	E	C	B	O	E	O	A
P	A	U	R	O	R	A	A	R	T	T	R	L
A	R	F	N	E	B	U	L	A	S	S	G	O
M	R	P	E	N	U	M	B	R	A	A	E	E

Just for Fun

Write an adventure story that takes place in outer space using ten or more of the words from the puzzle.

Name ______________________________ Date ______________

Weather Search

SPELLING AND SCIENCE VOCABULARY

Find and circle the weather words hidden in the puzzle. Words can be found going forward, backward, up, down, and diagonally.

blizzard	fog	hurricane	rain	thunder
cloud	frost	icicle	sleet	tornado
cumulus	hailstone	lightning	snow	typhoon
cyclone	humidity	monsoon	temperature	windstorm

F	R	O	S	T	E	E	L	S	I	H	U
T	Y	P	H	O	O	N	N	W	A	W	D
E	G	E	L	R	E	O	B	I	T	N	T
C	I	R	N	N	W	L	L	N	S	G	M
L	H	U	O	A	L	S	I	D	U	N	G
O	C	T	F	D	T	A	Z	S	L	I	E
U	R	A	C	O	R	Y	Z	T	U	N	L
D	O	R	N	Y	G	E	A	O	M	T	C
N	T	E	N	A	C	I	R	R	U	H	I
G	U	P	O	O	U	L	D	M	C	G	C
T	L	M	O	N	S	O	O	N	S	I	I
Y	R	E	D	N	U	H	T	N	O	L	A
G	Y	T	I	D	I	M	U	H	E	P	R

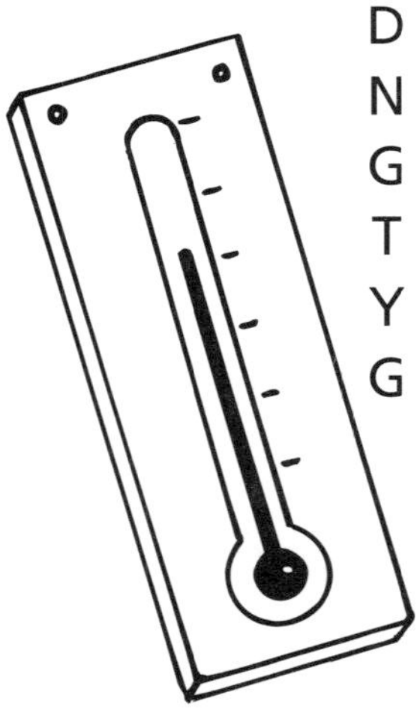

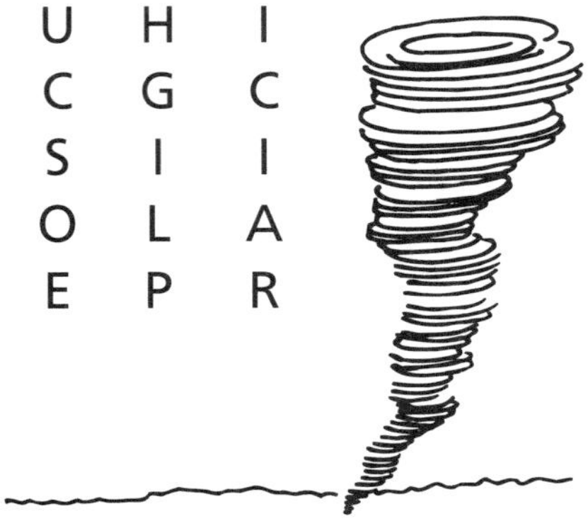

Just for Fun

Pretend you are a weather reporter. Write a weather forecast for the nation using at least six of the weather words from the puzzle.

Name ______________________________ Date ______________

Capital Search

SPELLING AND SOCIAL STUDIES VOCABULARY

Find and circle the United States capitals hidden in the puzzle. Words can be found going forward, backward, up, down, and diagonally.

Annapolis	Cheyenne	Hartford	Olympia	Sacramento
Atlanta	Concord	Juneau	Pierre	Springfield
Boise	Dover	Lansing	Providence	Tallahassee
Boston	Indianapolis	Montgomery	Raleigh	Trenton

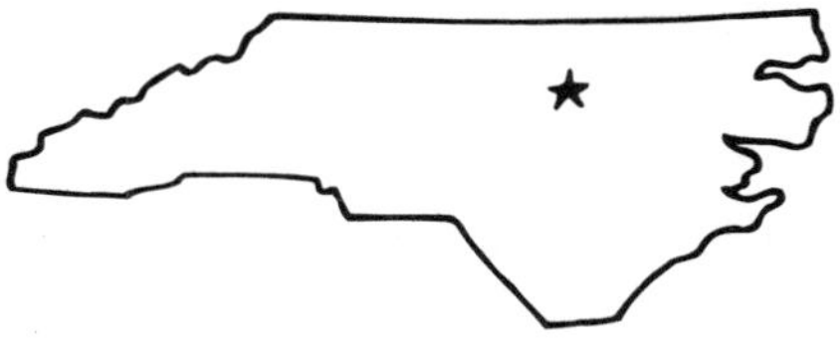

M	R	L	O	P	N	O	T	N	E	R	T	T	M
P	I	E	R	R	E	N	A	C	E	A	D	N	S
U	O	N	V	E	I	T	N	G	O	L	E	O	O
E	T	R	O	O	N	E	B	N	E	E	C	E	M
R	N	L	P	A	D	O	N	I	S	I	H	F	N
A	E	L	L	I	I	H	F	S	L	G	E	D	O
N	M	T	V	S	A	G	A	N	N	H	Y	R	C
N	A	O	E	S	N	H	A	A	O	T	E	O	S
A	R	C	S	I	A	I	I	L	T	O	N	F	O
P	C	Y	R	L	P	A	C	A	S	C	N	T	G
O	A	P	L	M	O	N	T	G	O	M	E	R	Y
L	S	A	Y	H	L	R	N	R	B	N	A	A	E
I	T	L	G	N	I	A	D	G	S	B	C	H	A
S	O	C	M	M	S	J	U	N	E	A	U	G	S

Just for Fun

List each capital from the puzzle. Then write the name of the state where each capital is located. Use a map if you need help.

Name ______________________________ Date ______________

Colonial Search

SPELLING AND SOCIAL STUDIES VOCABULARY

Find and circle the colonial words hidden in the puzzle. Words can be found going forward, backward, up, down, and diagonally.

blacksmith	flax	lantern	plantation	sconce
breeches	gristmill	meeting house	Puritans	spinning wheel
candles	hornbook	musket	quilting bee	tallow
colonists	indigo	pewter	Quakers	venison

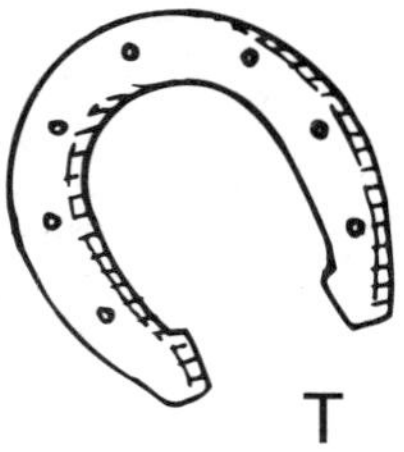

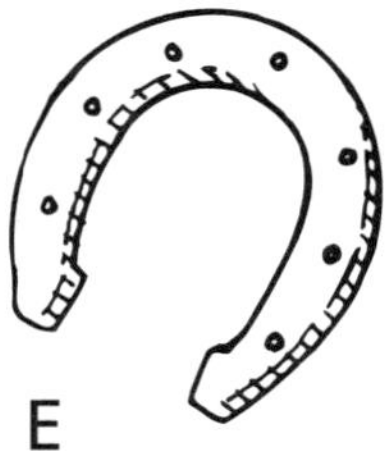

T	S	Q	B	L	H	H	M	U	S	K	E	T	E
O	N	U	L	X	L	U	O	I	W	S	H	I	C
R	O	I	A	E	S	I	M	R	U	I	L	P	N
V	O	L	C	C	P	E	M	O	N	L	S	I	O
E	F	T	K	A	B	U	H	T	G	B	E	N	C
N	C	I	S	T	L	G	R	C	S	I	O	D	S
I	A	N	M	T	N	C	K	I	E	I	D	O	A
S	N	G	I	I	U	L	A	N	T	E	R	N	K
O	D	B	T	W	O	L	L	A	T	A	R	G	I
N	L	E	H	P	E	W	T	E	R	P	N	B	O
L	E	E	H	W	G	N	I	N	N	I	P	S	O
M	S	R	E	K	A	U	Q	T	E	O	I	N	O
B	U	C	O	L	O	N	I	S	T	S	E	K	L
S	H	R	P	M	E	I	N	E	R	K	T	T	N

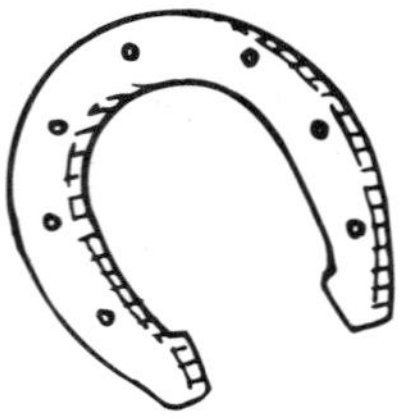

Just for Fun
List eight words from the puzzle. Write the definitions of the words, and use each word in a sentence. Use a dictionary if you need help.

Name ______________________________ Date ______________

Country Search

SPELLING AND SOCIAL STUDIES VOCABULARY

Find and circle the names of the countries hidden in the puzzle. Words can be found going forward, backward, up, down, and diagonally.

Angola	Chile	Gabon	Iraq	Kenya
Argentina	China	Guatemala	Israel	Mexico
Belgium	Denmark	Hungary	Jamaica	Panama
Brazil	France	India	Japan	Venezuela

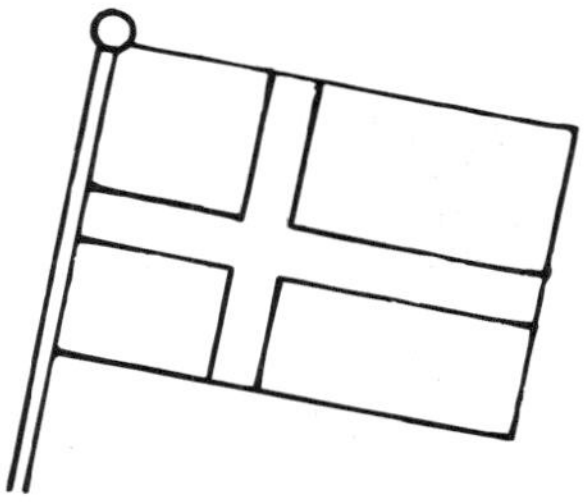

A	A	M	A	N	A	P	G	A	J	A	B
L	L	A	I	A	L	L	E	A	R	S	I
A	E	O	M	D	A	I	A	C	B	C	M
M	U	I	G	L	E	B	Z	I	R	O	I
E	Z	O	I	N	I	N	Z	A	T	Y	N
T	E	C	I	R	A	Q	M	M	R	X	D
A	N	I	T	N	E	G	R	A	N	B	I
U	E	X	E	G	A	N	G	J	R	A	A
G	V	E	F	R	A	N	C	E	Y	K	A
N	G	M	U	N	U	L	I	N	D	A	L
H	I	N	C	H	I	L	E	H	I	I	A
J	A	P	A	N	H	K	Z	B	C	I	C

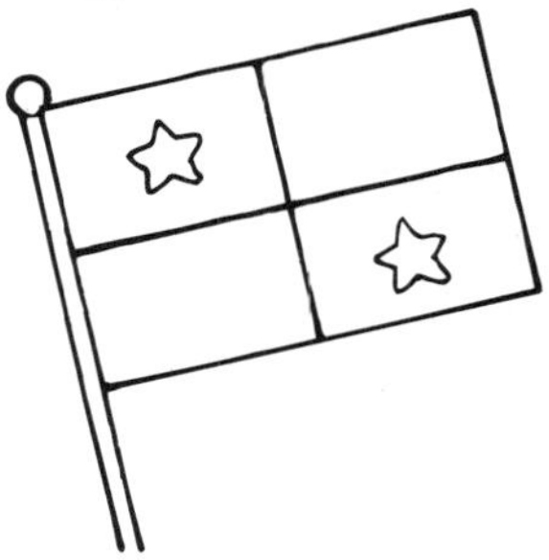

Just for Fun

Make a chart with the following headings: North America, South America, Africa, Asia, and Europe. Then list each country in the puzzle under the correct heading.

Name ______________________________ Date ______________

Geography Search

SPELLING AND SOCIAL STUDIES VOCABULARY

Find and circle the geography words hidden in the puzzle. Words can be found going forward, backward, up, down, and diagonally.

archipelago	canyon	fjord	mesa	plateau
basin	crevasse	gulf	mountain	tributary
bay	delta	island	pampa	veldt
bayou	desert	isthmus	peninsula	wetlands

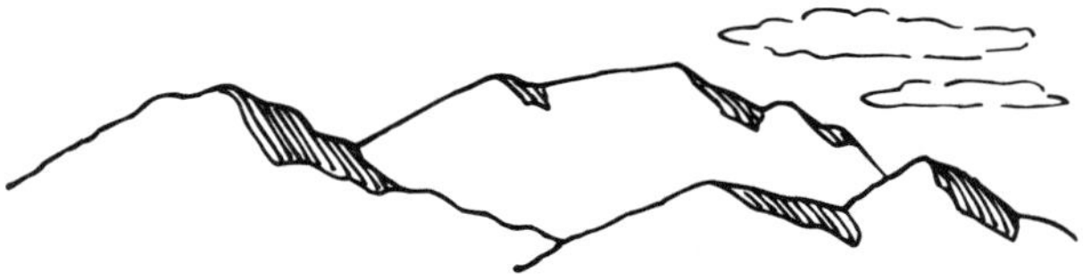

R	N	F	R	A	A	N	S	A	I	D	P
N	E	C	S	D	E	U	J	F	L	U	G
D	S	L	M	E	I	T	L	A	P	O	A
R	B	B	A	S	I	N	A	A	C	Y	D
O	G	A	L	E	P	I	H	C	R	A	A
J	S	A	Y	R	E	S	G	L	E	B	N
F	N	I	A	T	N	U	O	M	V	U	O
D	S	S	T	R	I	B	U	T	A	R	Y
W	E	T	L	A	N	D	S	E	S	I	N
M	A	H	E	N	S	I	T	L	S	R	A
A	A	M	D	Y	U	A	E	S	E	E	C
C	D	U	T	D	L	E	V	A	O	S	T
A	T	S	M	P	A	M	P	A	C	U	O

Just for Fun

List eight words from the puzzle. Write the definitions of the words, and use each word in a sentence. Use a dictionary if you need help.

Name ______________________________ Date ______________

Native American Search

SPELLING AND SOCIAL STUDIES VOCABULARY

Find and circle the names of the Native American tribes hidden in the puzzle. Words can be found going forward, backward, up, down, and diagonally.

Algonquin	Cheyenne	Dakota	Navajo	Shawnee
Apache	Chippewa	Hopi	Osage	Sioux
Arapaho	Choctaw	Kickapoo	Ottawa	Yuma
Cherokee	Crow	Mohegan	Pawnee	Zuni

I	O	E	E	N	W	A	P	E	A	C
A	P	A	D	A	K	O	T	A	Z	H
W	D	O	O	G	A	E	E	U	W	O
A	A	C	H	E	Y	E	N	N	E	C
X	R	T	R	H	K	I	E	Y	H	T
U	A	L	G	O	N	Q	U	I	N	A
O	P	Z	R	M	W	M	P	E	A	W
I	A	E	H	C	A	P	A	G	V	A
S	H	A	W	N	E	E	E	A	A	T
C	O	K	O	W	P	S	U	S	J	T
O	O	P	A	K	C	I	K	O	O	O
T	O	K	A	E	E	N	A	A	A	P

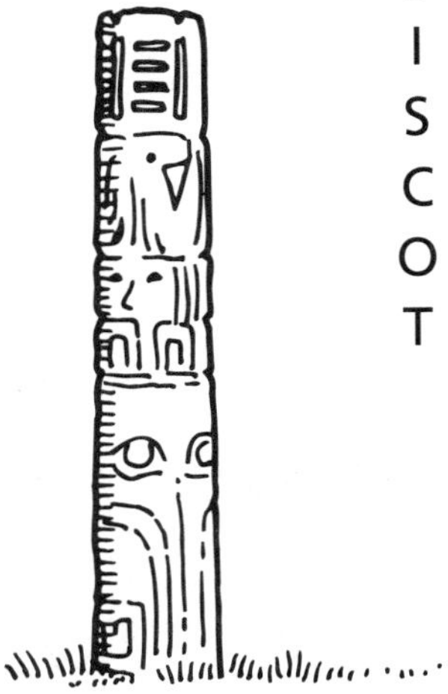

Just for Fun

Research ten of the tribes from the puzzle, and identify where each tribe settled. Make a bar or circle graph of your findings.

Name ______________________________ Date ______________

President Search

SPELLING AND SOCIAL STUDIES VOCABULARY

Find and circle the last names of the United States presidents hidden in the puzzle. Words can be found going forward, backward, up, down, and diagonally.

Adams (John Quincy)	Grant	Lincoln	Roosevelt (Franklin D.)
Bush (George Walker)	Kennedy	Madison	Taylor
Carter	Jackson	Monroe	Truman
Clinton	Jefferson	Nixon	Washington
Ford	Johnson (Lyndon B.)	Reagan	Wilson

N	A	M	U	R	T	M	F	N	T	S
O	C	L	I	N	T	O	N	R	N	M
S	A	I	O	K	R	N	O	T	I	A
I	R	N	N	D	E	L	T	L	X	D
D	T	C	J	O	Y	N	G	E	O	A
A	E	O	R	A	S	I	N	V	N	D
M	R	L	T	N	C	N	I	E	S	I
O	T	N	A	R	G	K	H	S	D	S
N	J	E	F	F	E	R	S	O	N	Y
R	E	A	G	A	N	E	A	O	J	E
O	R	N	O	S	L	I	W	R	N	N
E	B	U	S	H	E	S	S	A	Y	N

Just for Fun

Write the last name of each president from the puzzle. Then number the order in which each president served office starting with Washington. Use reference materials and the Internet for help.

Name ______________________________ Date ______________

State Search

Spelling and Social Studies Vocabulary

Find and circle the names of the states hidden in the puzzle. Words can be found going forward, backward, up, down, and diagonally.

Alabama	Iowa	Minnesota	Oklahoma	Virginia
Colorado	Indiana	Nebraska	Pennsylvania	Washington
Delaware	Kentucky	New Mexico	Texas	West Virginia
Florida	Massachusetts	North Carolina	Utah	Wyoming

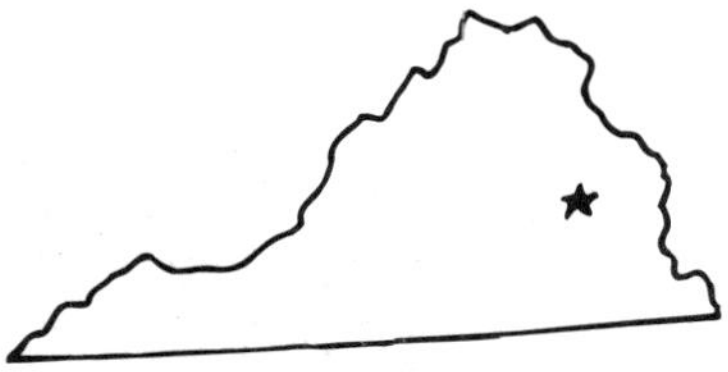

I	D	U	E	R	I	K	E	I	N	G	A	S	I
I	S	T	T	E	S	U	H	C	A	S	S	A	M
A	N	W	E	S	T	V	I	R	G	I	N	I	A
H	A	D	M	X	S	E	O	O	O	I	O	N	I
M	L	H	I	E	A	W	W	D	L	R	T	A	N
N	A	O	N	A	W	S	A	O	Y	R	G	V	I
I	B	A	N	Y	N	R	R	D	K	S	N	L	G
E	A	U	E	A	O	A	E	M	C	S	I	Y	R
X	M	S	S	L	C	L	A	M	U	N	H	S	I
C	A	M	O	H	A	L	K	O	T	I	S	N	V
Y	R	C	T	W	Y	O	M	I	N	G	A	N	O
E	A	R	A	U	O	C	I	X	E	M	W	E	N
I	O	R	E	T	T	O	D	N	K	O	U	P	A
N	E	B	R	A	S	K	A	D	I	R	O	L	F
W	G	A	W	H	N	S	S	A	G	A	A	N	E

Just for Fun

Write the name of each state from the puzzle in the order in which each was admitted to the Union. Use reference materials and the Internet for help.

Name ______________________________ Date ______________

City Search

Spelling and Social Studies Vocabulary

Find and circle the names of the U.S. cities hidden in the puzzle. Words can be found going forward, backward, up, down, and diagonally.

Albuquerque	Charlotte	Eugene	Miami	Tucson
Amarillo	Cincinnati	Knoxville	Norfolk	Tulsa
Burlington	Columbus	Los Angeles	Omaha	Wilmington
Casper	Detroit	Madison	Pueblo	Worcester

I	L	C	K	M	C	R	I	S	N	R	L	T
N	O	S	I	D	A	M	L	A	E	E	A	U
A	L	E	E	Q	A	U	I	N	D	P	N	L
S	L	A	R	I	B	B	E	E	N	S	O	S
W	I	L	M	I	N	G	T	O	N	A	S	A
K	R	B	R	N	U	R	T	O	E	C	C	L
L	A	U	N	E	O	G	E	R	M	H	U	O
O	M	Q	C	I	N	C	I	N	N	A	T	I
F	A	U	T	I	N	E	R	U	P	R	H	U
R	S	E	L	E	G	N	A	S	O	L	O	A
O	N	R	M	E	L	L	I	V	X	O	N	K
N	U	Q	C	W	O	R	C	E	S	T	E	R
B	S	U	B	M	U	L	O	C	X	T	M	C
P	U	E	B	L	O	A	O	G	R	E	N	N

Just for Fun

Make a word search (with an answer key) that includes the names of 20 cities located in the state in which you live. Give your puzzle to a classmate to complete.

Name ______________________ Date ______________

Parts of Speech

VOCABULARY

For each category write a word that begins with the letter on the left. Score one point for each correct answer. Earn five bonus points for any category where you have no incorrect answers or blanks. Use a dictionary if you need help finding a word for any category.

Letter	Noun	Verb	Adjective	Adverb	Preposition	Score
E						
O						
T						
I						
B						
					Total	
					Bonus	
					Final Score	

Name ______________________ Date ______________

Living Things

VOCABULARY

For each category write a word that begins with the letter on the left. Score one point for each correct answer. Earn five bonus points for any category where you have no incorrect answers or blanks. Use books, encyclopedias, and other reference materials if you need help finding a word for any category.

Letter	Mammal	Bird	Fish	Flower	Tree	Score
R						
A						
P						
M						
S						
					Total	
					Bonus	
					Final Score	

Name ______________________ Date ______________

The World of Science

VOCABULARY

For each category write a word that begins with the letter on the left. Score one point for each correct answer. Earn five bonus points for any category where you have no incorrect answers or blanks. Use books, encyclopedias, and other reference materials if you need help finding a word for any category.

Letter	Weather Word	Part of the Body	Invention	Computer Term	Solar System Word	Score
E						
S						
N						
B						
A						
					Total	
					Bonus	
					Final Score	

Name ____________________ Date ____________

Geography

VOCABULARY

For each category write a word that begins with the letter on the left. Score one point for each correct answer. Earn five bonus points for any category where you have no incorrect answers or blanks. Use maps, globes, atlases, encyclopedias, and other reference materials if you need help finding a word for any category.

Letter	City	Country	Body of Water (ocean, river, lake)	Mountain	Island	Score
A						
N						
F						
K						
E						
					Total	
					Bonus	
					Final Score	

Name ______________________ Date ______________

Places

VOCABULARY

For each category write a word that begins with the letter on the left. Score one point for each correct answer. Earn five bonus points for any category where you have no incorrect answers or blanks. Use maps, globes, atlases, encyclopedias, and other reference materials if you need help finding a word for any category.

Letter	State	Foreign Country	City in Your State	Tourist Attraction	U.S. Capital City	Score
A						
N						
P						
F						
G						
					Total	
					Bonus	
					Final Score	

Name ______________________ Date ______________

People

VOCABULARY

For each category write a word that begins with the letter on the left. Score one point for each correct answer. Earn five bonus points for any category where you have no incorrect answers or blanks. Use books, encyclopedias, and other reference materials if you need help finding a word for any category.

Letter	Author (last name)	Boy's Name	Girl's Name	Famous Person (last name)	Occupation	Score
P						
O						
R						
J						
E						
					Total	
					Bonus	
					Final Score	

Name ______________________ Date ______________

This and That

VOCABULARY

For each category write a word that begins with the letter on the left. Score one point for each correct answer. Earn five bonus points for any category where you have no incorrect answers or blanks. Use books, magazines, and other reference materials if you need help finding a word for any category.

Letter	Television Show	Sport or Game	Book Title (skip *a, an, the*)	Color	Movie Title (skip *a, an, the*)	Score
B						
G						
R						
N						
P						
					Total	
					Bonus	
					Final Score	

Name ______________________ Date ______________________

Food Fun

VOCABULARY

For each category write a word that begins with the letter on the left. Score one point for each correct answer. Earn five bonus points for any category where you have no incorrect answers or blanks. Use books, encyclopedias, and other reference materials if you need help finding a word for any category.

Letter	Fruit	Vegetable	Dessert	Main Course	Drink	Score
A						
T						
C						
B						
O						
					Total	
					Bonus	
					Final Score	

Name ______________________ Date ______________

Let's Go Shopping

VOCABULARY

For each category write a word that describes something you would buy at that store. The words should begin with the letter on the left. Score one point for each correct answer. Earn five bonus points for any category where you have no incorrect answers or blanks. Use newspapers, magazines, and other reference materials if you need help finding a word for any category.

Letter	Department Store	Hardware Store	Office Supply Store	Toy Store	Grocery Store	Score
B						
C						
H						
L						
T						
					Total	
					Bonus	
					Final Score	

Name ______________________ Date ______________

Fun Words

VOCABULARY

For each category write a word that begins with the letter on the left. Score one point for each correct answer. Earn five bonus points for any category where you have no incorrect answers or blanks. Use books, encyclopedias, and other reference materials if you need help finding a word for any category.

Letter	Something Round	Form of Transportation	Song Title	A Game to Play	Something Made of Wood	Score
S						
P						
H						
W						
T						
					Total	
					Bonus	
					Final Score	

Name ______________________ Date ____________

Synonym Puzzle

SYNONYMS, VOCABULARY, SPELLING

A **synonym** is a word that means the same, or almost the same, as another word. For example, *false* and *untrue* are synonyms.

Write a synonym for each word to complete the crossword puzzle.

Across
2. fast
4. easy
6. late
8. vanish
10. close
12. silly
14. mend

Down
1. shake
3. small
5. reply
7. buy
9. vacant
11. smell
13. monkeys

Just for Fun
Write a synonym for each word.

occur ____________ rage ____________

quick ____________ timid ____________

sincere ____________

Name ______________________________ Date ______________

Antonym Puzzle

ANTONYMS, VOCABULARY, SPELLING

An **antonym** is a word that means the opposite of another word. For example, *dark* is an antonym of *light*.

Write an antonym for each word to complete the crossword puzzle.

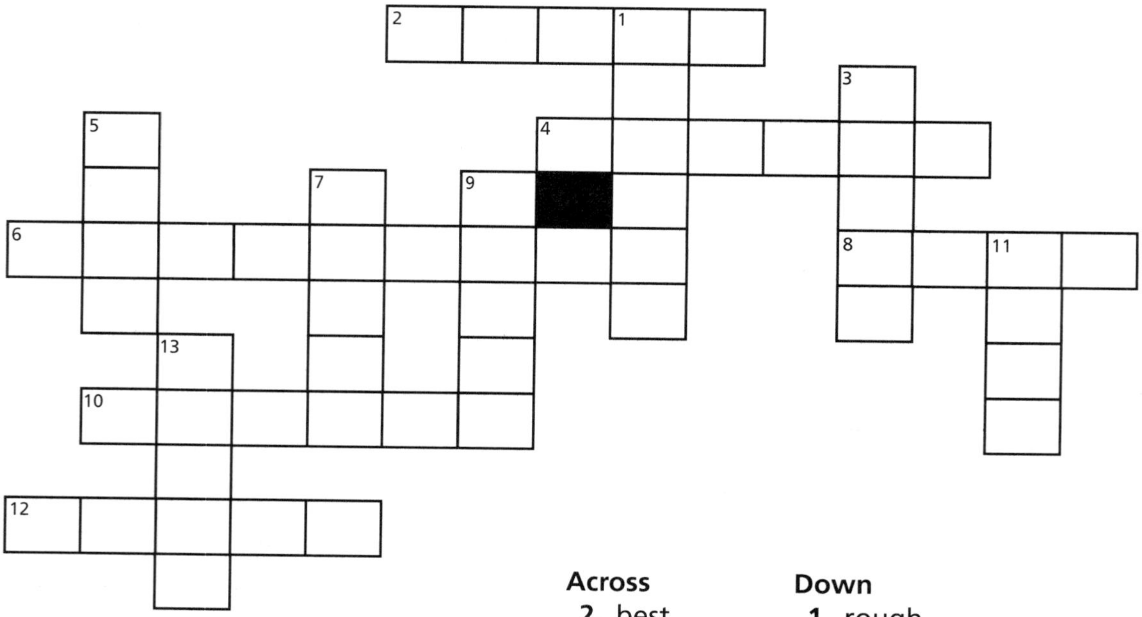

Across
2. best
4. allow
6. alike
8. wild
10. wide
12. day

Down
1. rough
3. clean
5. achieve
7. always
9. above
11. few
13. small

Just for Fun
Write an antonym for each word.

stay ______________________ crooked ______________________

shallow ______________________ innocent ______________________

healthy ______________________

Name ____________________ Date ____________

Homophone Puzzle

HOMOPHONES, VOCABULARY, SPELLING

A **homophone** is a word that sounds like another word (or words) but has a different meaning and spelling. For example, *to*, *too*, and *two* are homophones .

Write a homophone for each word to complete the crossword puzzle.

Across
2. capitol
4. flour
6. bred
8. need
10. aloud
12. here
13. peace

Down
1. hangar
3. blue
5. isle
7. doe
8. new
9. right
11. hire

Just for Fun
Write a homophone for each word.

stair ____________________

poll ____________________

weighed ____________________

principal ____________________

week ____________________

Name ______________________________ Date ____________

Past Participle Puzzle

VERB TENSES, VOCABULARY, SPELLING

The **past participle** of a verb shows that the verb's action took place in the past. For example, the past participle of the verb *see* (*I see her each day.*) is *seen* (*I have seen her before.*). The helping verbs *has* and *have* are always used with the past participle.

Write the past participle for each verb to complete the crossword puzzle.

Across
2. rise
4. do
6. break
8. begin
10. throw
12. write
13. sink

Down
1. steal
3. choose
5. forget
7. grow
9. swim
11. bite

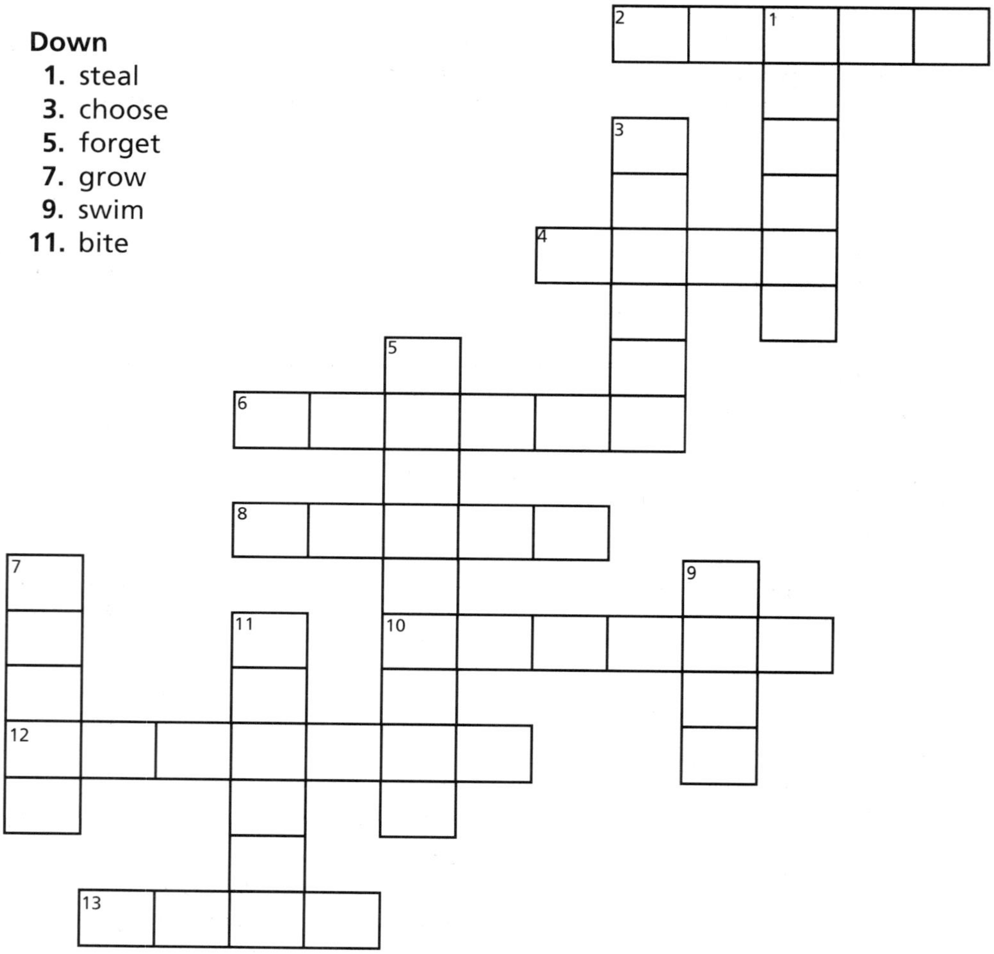

Just for Fun

Write the past participle form of each verb.

draw ______________ run ______________

go ______________ be ______________

speak ______________

Name ______________________ Date ____________

Noun Puzzle

NOUNS, VOCABULARY, SPELLING

A **noun** is the name of a person, place, thing, quality, or idea.

Write the noun that best matches each definition to complete the crossword puzzle.

brink	neighbor	periscope	speck
knuckle	coward	pioneer	teen
orbit	path	progress	wish

Across

2. a small spot or mark
4. a person who goes before, opening the way for others to follow
6. a joint of the finger
8. a longing or desire for something
10. a track made by or for walking
12. path followed by a moon or planet

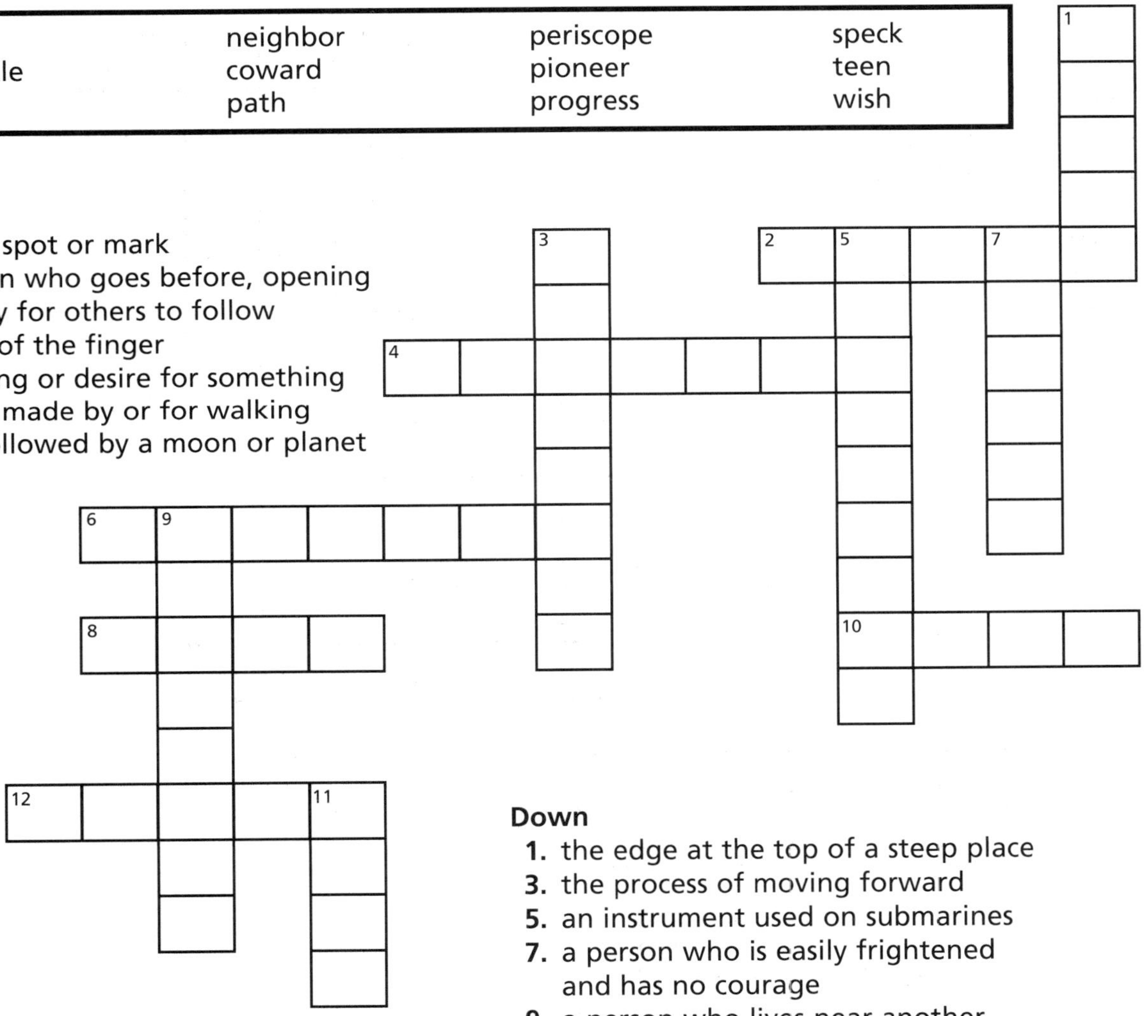

Down

1. the edge at the top of a steep place
3. the process of moving forward
5. an instrument used on submarines
7. a person who is easily frightened and has no courage
9. a person who lives near another
11. the years from 13 through 19

Just for Fun

Write the definition of each noun.

ballad ______________________

ballot ______________________

barge ______________________

Name ______________________ Date __________

Plural Puzzle

NOUNS, PLURALS, SPELLING

A **noun** is the name of a person, place, thing, quality, or idea. A noun that is *singular* names only one person, place, thing, quality, or idea. A noun that is *plural* names more than one.

Write the plural form of each singular noun to complete the crossword puzzle.

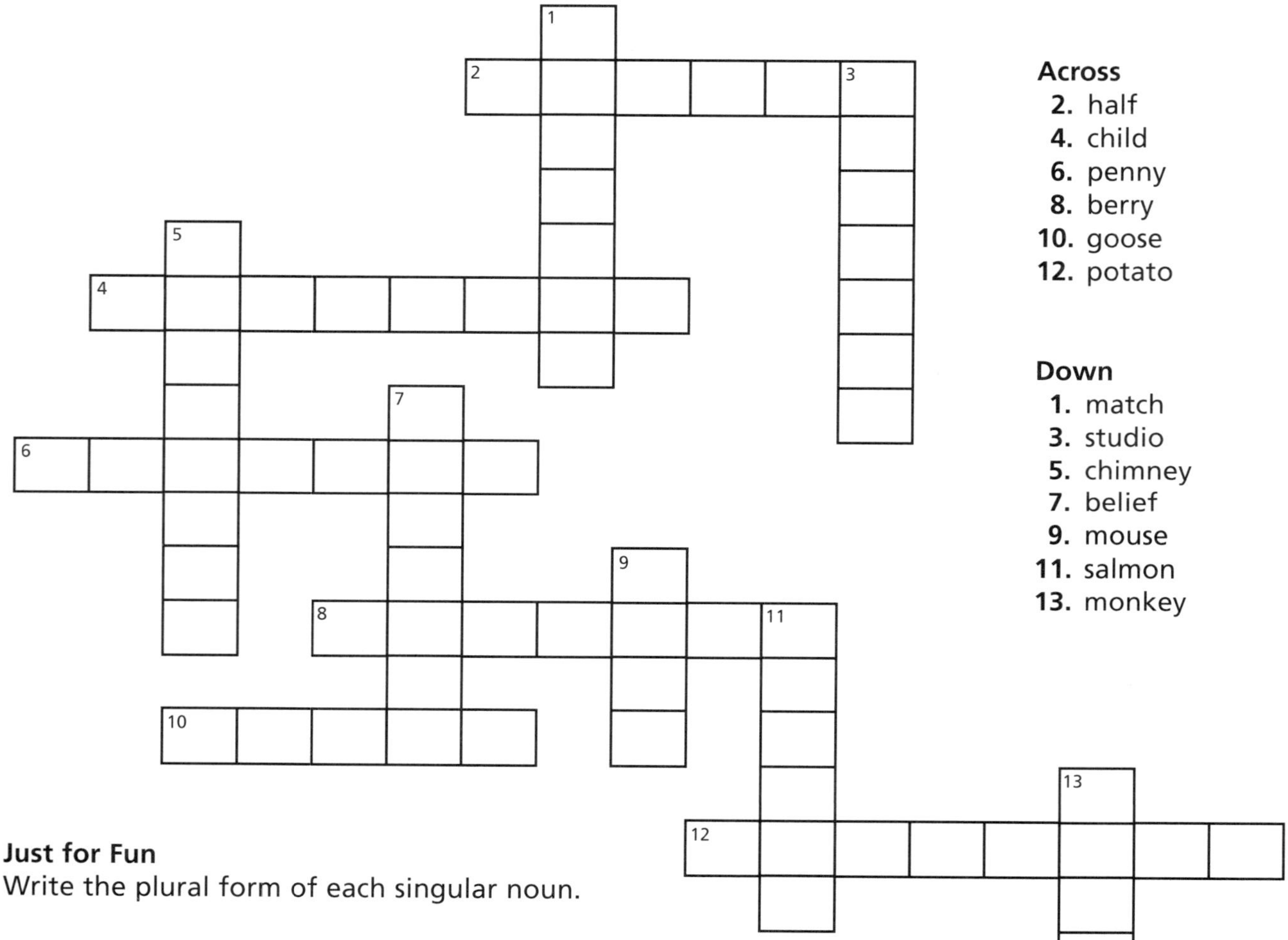

Across
2. half
4. child
6. penny
8. berry
10. goose
12. potato

Down
1. match
3. studio
5. chimney
7. belief
9. mouse
11. salmon
13. monkey

Just for Fun
Write the plural form of each singular noun.

pencil ______________________

sandwich ______________________

sheep ______________________

foot ______________________

calf ______________________

Name ______________________________ Date ______________

Adjective Puzzle

ADJECTIVES, VOCABULARY, SPELLING

An **adjective** is a word that describes a noun or pronoun. An adjective tells how many, what kind, or which one.

Write the adjective that best matches each definition to complete the crossword puzzle.

afraid	glossy	firm	hectic
cruel	circular	various	obvious
eager	lazy	grumpy	phony

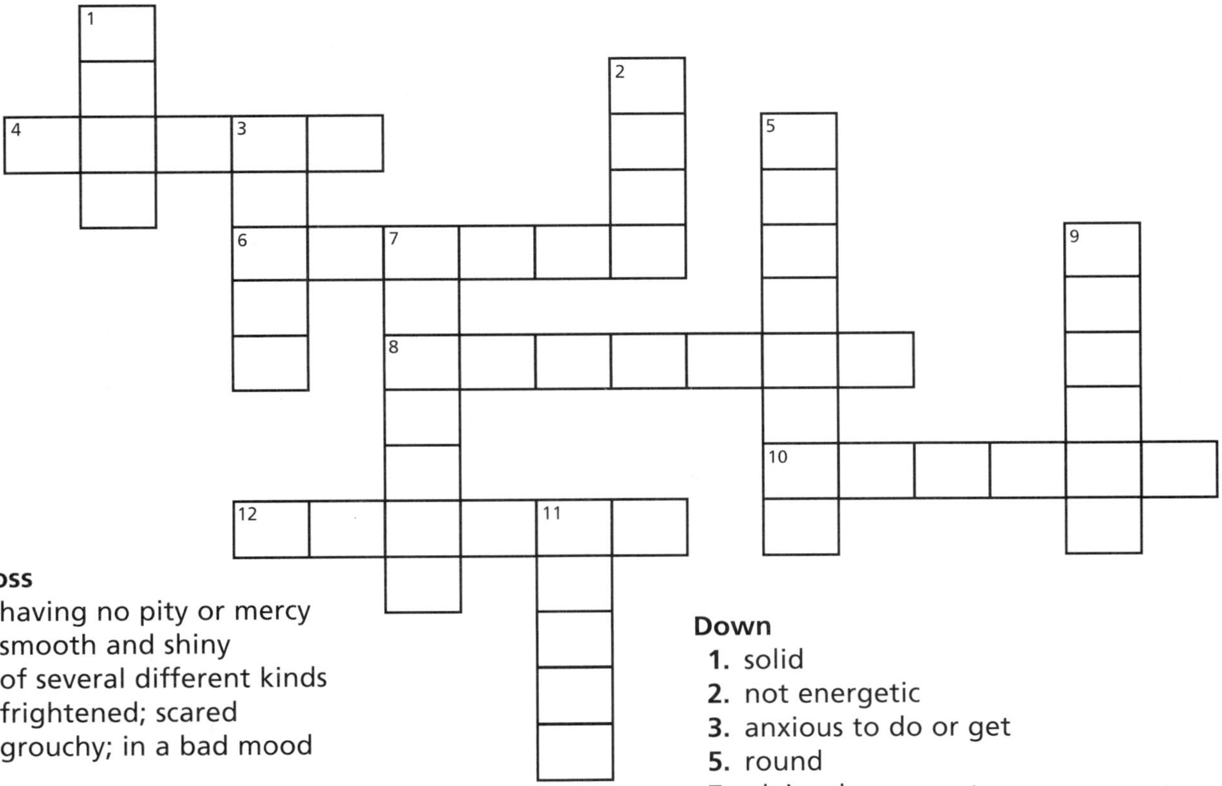

Across

4. having no pity or mercy
6. smooth and shiny
8. of several different kinds
10. frightened; scared
12. grouchy; in a bad mood

Down

1. solid
2. not energetic
3. anxious to do or get
5. round
7. plain; clear; easy to see
9. full of confusion or excitement
11. false; not real

Just for Fun
Write the definition of each adjective.

selfish ______________________________

secure ______________________________

similar ______________________________

Name ____________________ Date __________

Abbreviation Puzzle

ABBREVIATIONS, MEASUREMENT, SPELLING

An **abbreviation** is a shortened or contracted form of a word or phrase such as *adj.* for *adjective* or *ave.* for *avenue*.

Write out the unit of measure for each abbreviation to complete the crossword puzzle. (Note: Use the singular form of the abbreviation.)

Across
2. ft.
5. tbsp.
6. gal.
8. pt.
10. in.
11. oz.
12. qt.

Down
1. doz.
3. lb.
4. mi.
5. tsp.
7. hr.
9. yr.

Just for Fun
Write the abbreviation for each word.

department ____________________ boulevard ____________________

miscellaneous ____________________ avenue ____________________

corporation ____________________

Name ______________________________ Date ______________

Mammal Puzzle

SCIENCE, VOCABULARY, SPELLING

Write the name of the mammal that best matches each description to complete the crossword puzzle.

bear	porcupine	kangaroo	leopard
cheetah	camel	tiger	panda
giraffe	elephant	gorilla	

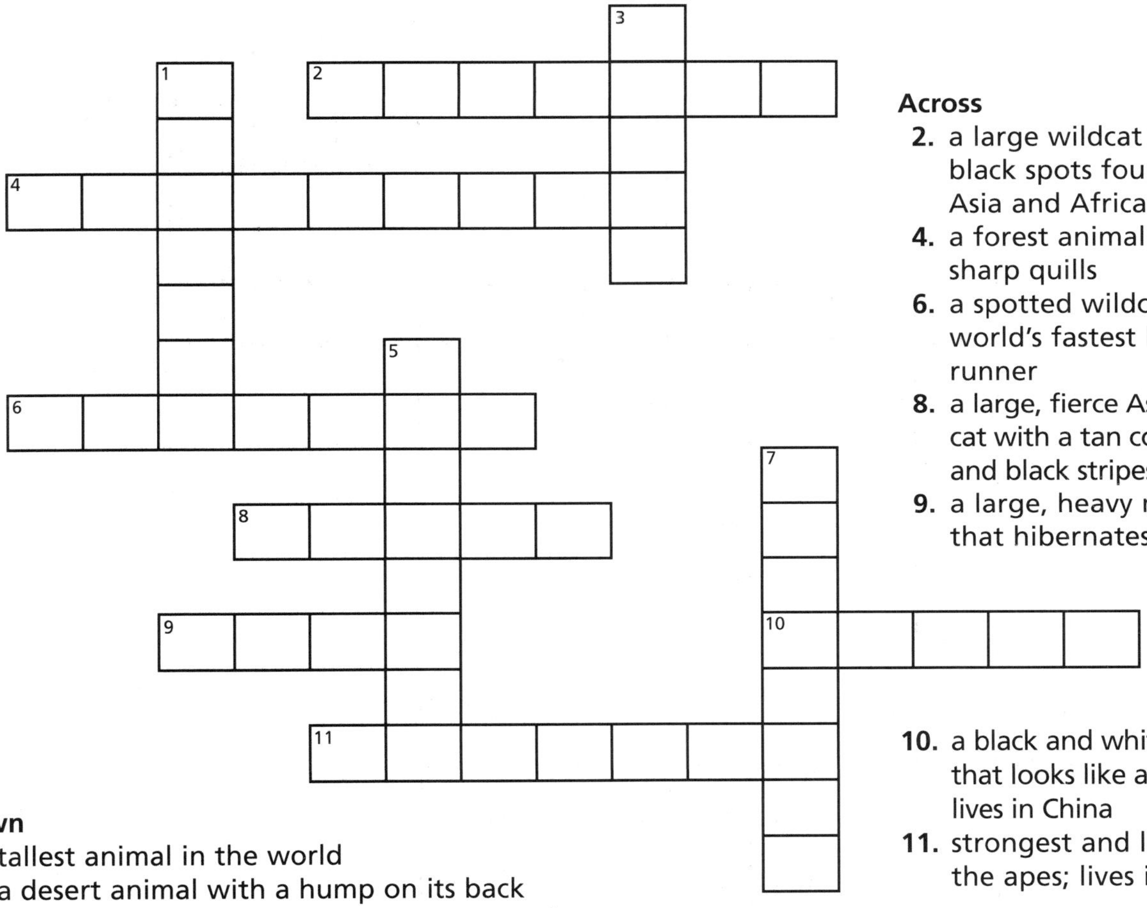

Across

2. a large wildcat with black spots found in Asia and Africa
4. a forest animal with sharp quills
6. a spotted wildcat; world's fastest land runner
8. a large, fierce Asian cat with a tan coat and black stripes
9. a large, heavy mammal that hibernates
10. a black and white animal that looks like a bear and lives in China
11. strongest and largest of the apes; lives in Africa

Down

1. tallest animal in the world
3. a desert animal with a hump on its back
5. a mammal that carries its young in a pouch
7. an animal with a trunk

Just for Fun

Make a word search (with an answer key) that includes the following names of mammals. Hide your words going up, down, backward, forward, and diagonally. Give your puzzle to a classmate to complete.

bat	baboon	coyote	eland
lemur	lion	mole	zebra

Name ______________________________ Date ______________

Animal Offspring Puzzle

SCIENCE, VOCABULARY, SPELLING

Write the name of the offspring for each animal to complete the crossword puzzle. Use an encyclopedia and other reference materials if you need help.

Across
2. deer
4. goat
6. horse
8. chicken
10. sheep
11. goose
12. kangaroo

Down
1. eagle
3. duck
5. cat
7. giraffe
8. bear
9. pig

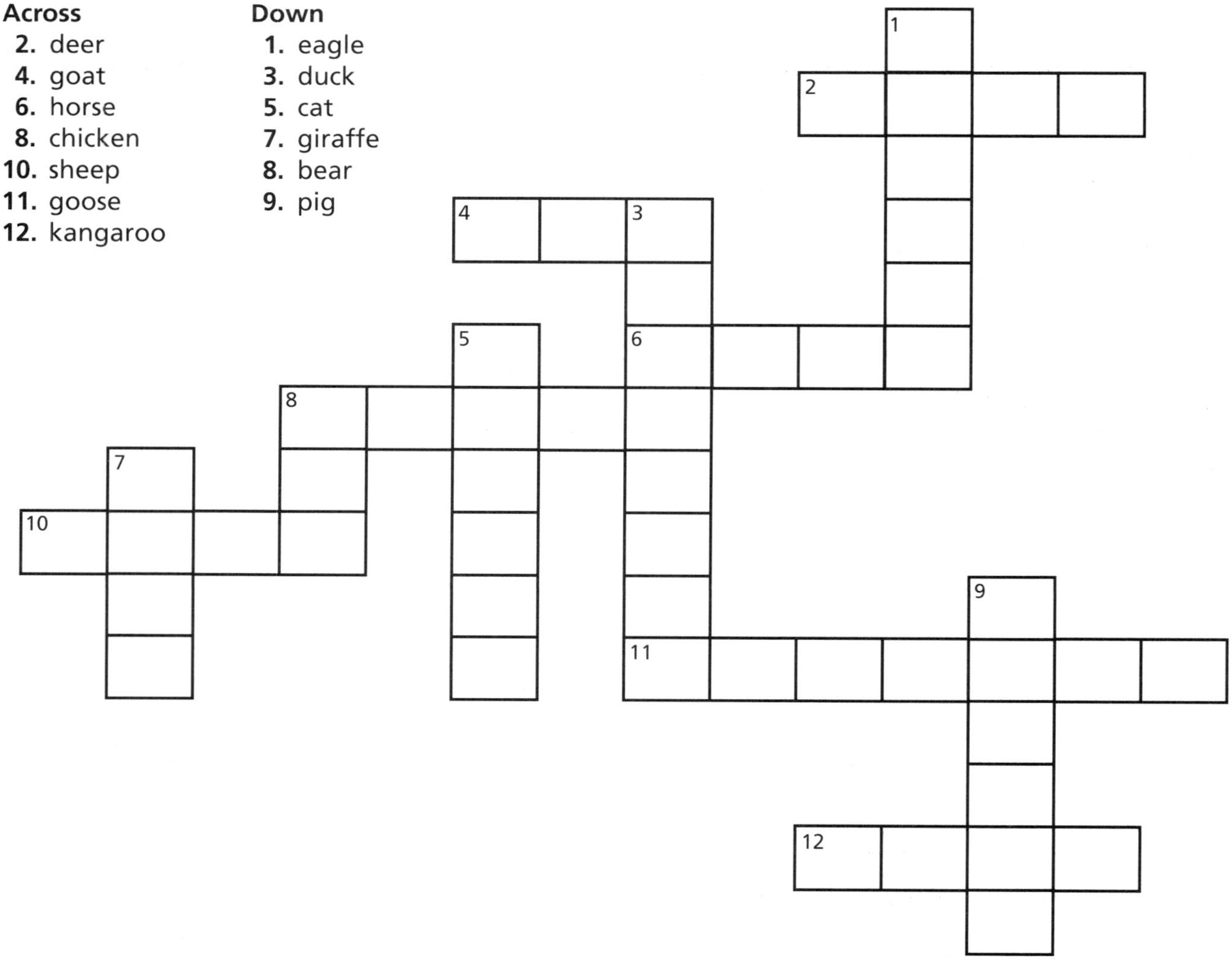

Just for Fun
Write the name of the offspring for each animal. Use an encyclopedia and other reference materials if you need help.

bobcat ______________________ whale ______________________

lion ______________________ antelope ______________________

zebra ______________________

Name ______________________________ Date ______________

Plant Puzzle

SCIENCE, VOCABULARY, SPELLING

Write the plant term that matches each description to complete the crossword puzzle.

botanist	root	redwoods	nectar
chlorophyll	cactus	stem	petals
evergreens	pollen	leaf	stigma

Across

2. the yellow powder on a flower's stamen
4. trees with leaves or needles that stay green all year
6. a person who studies plants
8. pigment that gives plants their green color
10. the part of the plant that receives pollen
12. a desert plant

Down

1. the tallest trees in the world
3. the colored leaf-like parts that make up the flower of a plant
5. the part of a plant that anchors it to the ground
7. sweet juice some plants make to attract insects
9. the flat, green part growing from the stem of a plant or tree
11. the part of the plant that carries food and water from the roots to the leaves

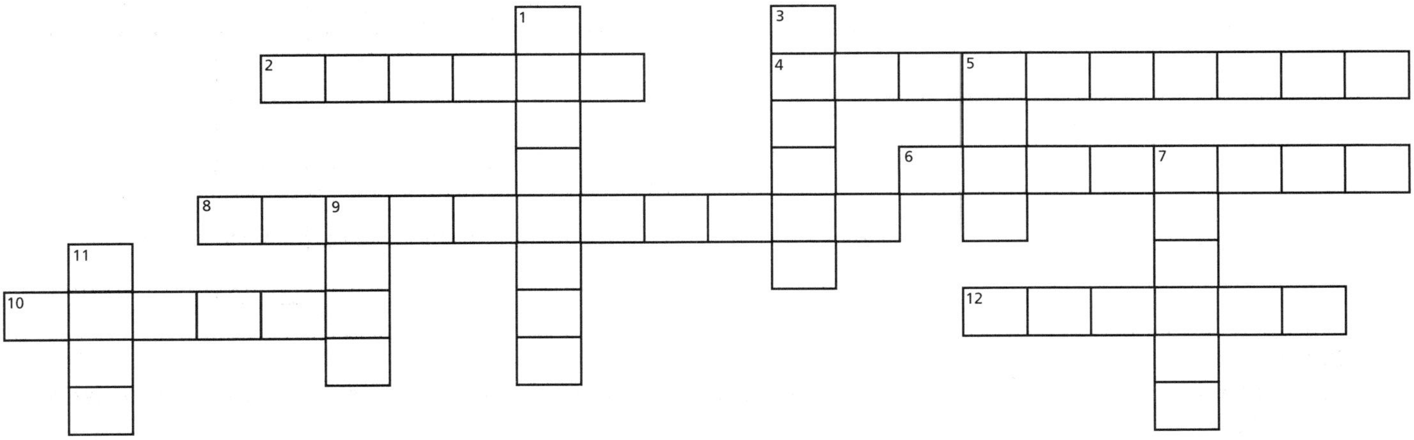

Just for Fun
Make a word search (with an answer key) that includes the twelve plant words from above. Hide your words going up, down, backward, forward, and diagonally. Give your puzzle to a classmate to complete.

Name ______________________________ Date ______________

Insect Puzzle

SCIENCE, VOCABULARY, SPELLING

Write the insect term that matches each description to complete the crossword puzzle.

bee	hive	flea	six
colony	caterpillar	termite	swarm
feelers	cricket	larva	two

Across

1. a group of ants living together
3. another name for an insect's antennae
5. an insect that makes a chirping sound
7. a wormlike animal that hatches from an insect's egg
8. number of legs an insect has
9. a group of bees
10. an insect that stings

Down

1. the larva of a butterfly or moth
2. the nest or home of bees
3. a small, biting insect that cannot fly but can jump far
4. an insect that eats wood
6. the number of wings a fly has

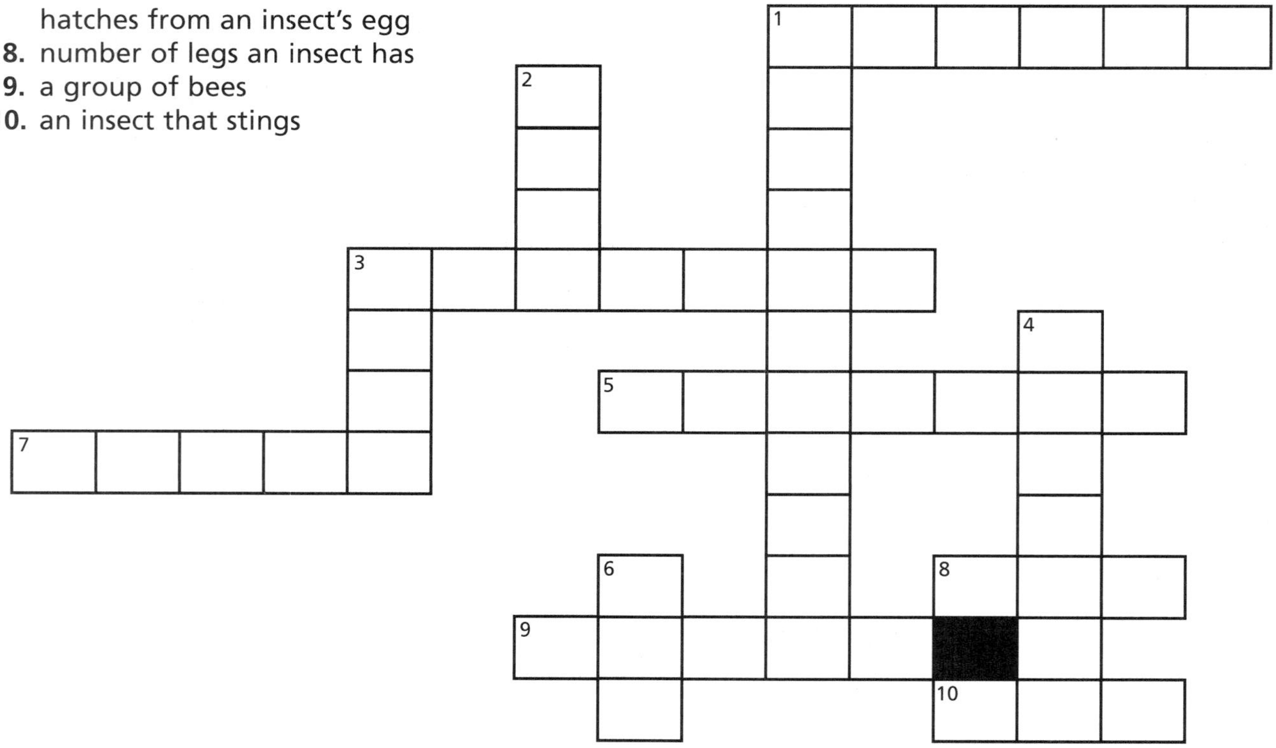

Just for Fun

Do research and learn more about one of these insects: katydid, grasshopper, beetle, dragonfly, fire ant, or wasp. Write ten fascinating facts about the insect you selected, and share your findings with your classmates in an oral report along with visual aids.

Name ______________________________ Date ______________

Bird Puzzle

SCIENCE, VOCABULARY, SPELLING

Write the bird term that matches each description to complete the crossword puzzle.

bald eagle	feathers	pink	peacock
beak	nest	rooster	red
bird	penguin	parrot	

Across

2. what a bird builds for laying its eggs and raising its young
4. the color of most flamingoes
6. these cover a bird, keep it warm, and help it to fly
7. a bird that can imitate speech and other sounds
8. the national bird of the U.S.
9. a warm-blooded vertebrate with feathers and wings

Down

1. an adult male chicken
3. the mouth of a bird
4. a bird with long green and gold feathers that spread like a fan
5. the color of most male cardinals
7. a black and white bird from Antarctica that cannot fly

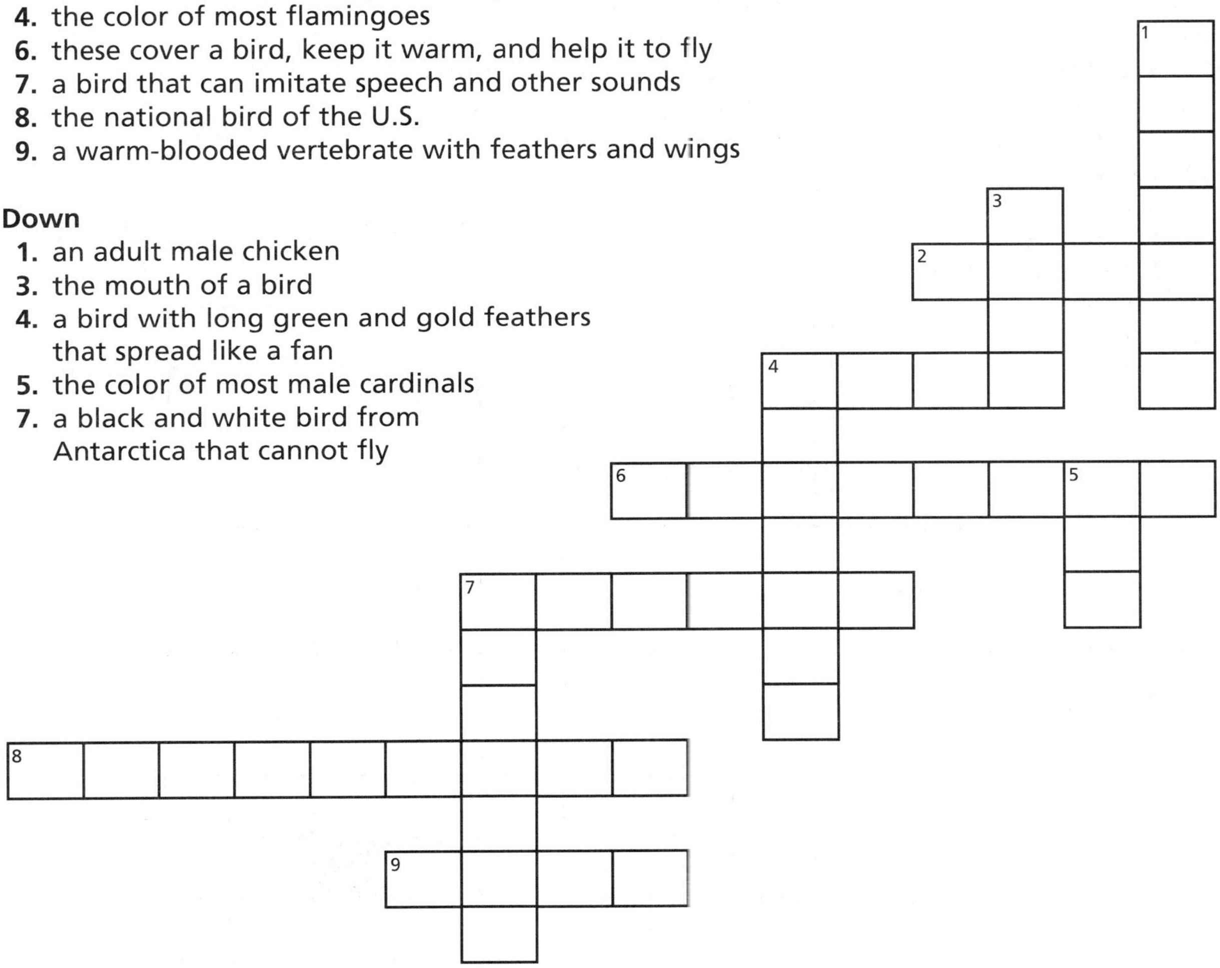

Just for Fun

Make a bird word search (with an answer key) that includes the names of these birds: condor, dove, emu, hawk, macaw, oriole, ostrich, roadrunner, woodpecker, quail, raven, crow, canary, hummingbird, heron, gull, pigeon, robin, ibis, and vulture. Hide your words going up, down, backward, forward, and diagonally. Give your puzzle to a classmate to complete.

Name ______________________________ Date ______________

Number, Please Puzzle

WEIGHT, MEASUREMENT, TIME, VOCABULARY, SPELLING

Write out the number for each description to complete the crossword puzzle. Some numbers may be used more than once. Use reference materials if you need help.

Across

2. number of seconds in a minute
4. number of ounces in a pound
6. number of months in a year
8. number of quarts in a gallon
10. number of inches in a foot

Down

1. number of years in a decade
3. number of feet in a yard
5. number of days in a week
6. number of teaspoons in a tablespoon
7. number of pints in a quart
9. number of years in a score

Just for Fun

Write the missing number for each equivalency.

_____ teaspoons = 1 tablespoon

_____ years = 1 century

_____ cups = 1 pint

_____ pounds = 1 ton

_____ hours = 1 day

Name ______________________________ Date ______________

What Would You Do with a Ballad?

VOCABULARY DEFINITIONS

What would you do with each of the following things whose names are printed in bold? Find and circle the *best* answer in each group of words. Use a dictionary if you need help.

1. What would you do with a **ballad**?
   - **a.** pry it open
   - **b.** sing it
   - **c.** fly it
   - **d.** give it a haircut

2. What would you do with a **badge**?
   - **a.** sit on it in your family room
   - **b.** plant it in a garden
   - **c.** get a cage for it
   - **d.** wear it

3. What would you do with a **balcony**?
   - **a.** fry it and eat it with eggs
   - **b.** rinse it with cold water
   - **c.** buy it in a grocery store
   - **d.** sit in it to watch a movie

4. What would you do with a **bandanna**?
   - **a.** feed it to a monkey
   - **b.** wear it on your head or around your neck
   - **c.** listen to it in an orchestra
   - **d.** use it to dig a hole

5. What would you do with a **barricade**?
   - **a.** catch it and eat it for dinner
   - **b.** plant it
   - **c.** take a detour around it
   - **d.** see it in an aquarium

6. What would you do with a **basin**?
   - **a.** fill it with water
   - **b.** chop it up for a salad
   - **c.** wear it on your feet
   - **d.** wear it on your head

7. What would you do with a **bassoon**?
   - **a.** visit it in a zoo
   - **b.** watch it on a circus trapeze
   - **c.** eat it for dessert
   - **d.** play it in a band

8. What would you do with **binoculars**?
   - **a.** wash the car with them
   - **b.** go bird watching with them
   - **c.** take photographs with them
   - **d.** serve them for dinner

9. What would you do with a **blazer**?
   - **a.** report it to the fire department
   - **b.** pour water on it
   - **c.** wear it with a pair of slacks
   - **d.** take a ride in it

10. What would you do with a **broth**?
    - **a.** eat it for dinner
    - **b.** sew it
    - **c.** buckle it up
    - **d.** paint with it

Name ______________________________ Date ______________

What Would You Do with a Cashew?

VOCABULARY DEFINITIONS

What would you do with each of the following things whose names are printed in bold? Find and circle the *best* answer in each group of words. Use a dictionary if you need help.

1. What would you do with a **cashew**?
   - **a.** put it in a cash register
   - **b.** spend it in a department store
   - **c.** eat it for a snack
   - **d.** blow your nose with it

2. What would you do with a **chameleon**?
   - **a.** watch it change colors
   - **b.** frame it
   - **c.** wear it around your shoulders
   - **d.** bake it

3. What would you do with a **chandelier**?
   - **a.** strum it
   - **b.** hang it on the ceiling for light
   - **c.** march to it
   - **d.** ride in it to a mountain top

4. What would you do with a **cinder**?
   - **a.** put it on a hot dog
   - **b.** drink it on a cold winter day
   - **c.** listen to it on the radio
   - **d.** sweep it up and throw it away

5. What would you do with **clogs**?
   - **a.** feed them corn in a trough
   - **b.** wear them on your feet
   - **c.** throw them in the dishwasher
   - **d.** watch them hanging from trees

6. What would you do with a **cockatoo**?
   - **a.** visit it at a zoo
   - **b.** play it in a rock band
   - **c.** drink it before dinner
   - **d.** read it

7. What would you do with a **container**?
   - **a.** use it to fire an employee
   - **b.** wear it on your teeth
   - **c.** tell it to a judge
   - **d.** put something inside it

8. What would you do with a **crepe**?
   - **a.** write on it
   - **b.** button it up
   - **c.** eat it
   - **d.** cut it out

9. What would you do with **currency**?
   - **a.** use it to buy things
   - **b.** pour water on it
   - **c.** swim against it in the ocean
   - **d.** bow to it

10. What would you do with **clutter**?
    - **a.** bathe it
    - **b.** take it for a walk for exercise
    - **c.** sharpen it
    - **d.** clean it up

Name ______________________________ Date ______________

What Would You Do with Galoshes?

Vocabulary Definitions

What would you do with each of the following things whose names are printed in bold? Find and circle the *best* answer in each group of words. Use a dictionary if you need help.

1. What would you do with **galoshes**?
   - **a.** eat them for dinner
   - **b.** store candy in them
   - **c.** fry them with butter
   - **d.** wear them over your shoes on a rainy day

2. What would you do with a **garnish**?
   - **a.** wear it on your face for a party
   - **b.** decorate food with it
   - **c.** use it to build a house
   - **d.** hammer it

3. What would you do with a **gauge**?
   - **a.** wrap it around a sore or wound
   - **b.** measure with it
   - **c.** look at it under a microscope
   - **d.** use it to dig a hole

4. What would you do with a **gerbil**?
   - **a.** measure it with a ruler
   - **b.** put it in an envelope and mail it
   - **c.** listen to it in an orchestra
   - **d.** keep it as a pet

5. What would you do with a **geyser**?
   - **a.** watch it shoot into the air
   - **b.** see it at a circus or carnival
   - **c.** tame it
   - **d.** drink it

6. What would you do with **ginger**?
   - **a.** paint a house with it
   - **b.** write a letter with it
   - **c.** add it to food for flavor
   - **d.** play it on a rainy day

7. What would you do with a **gladiolus**?
   - **a.** watch it fight in an arena
   - **b.** put it in a vase
   - **c.** strum it like a harp
   - **d.** put a coat of glaze on it

8. What would you do with a **gnu**?
   - **a.** carry it in your coat pocket
   - **b.** sharpen it
   - **c.** put it on a bagel with cream cheese
   - **d.** see it in Africa or the zoo

9. What would you do with a **gondola**?
   - **a.** take a ride on a canal in it
   - **b.** eat it with a ham sandwich
   - **c.** fly to Europe on it
   - **d.** decorate a hat with it

10. What would you do with a **guava**?
    - **a.** put it in a cage
    - **b.** spray your hair with it to keep it in place
    - **c.** eat it
    - **d.** pet it

Name ______________________________ Date ____________

What Would You Do with a Javelin?

VOCABULARY DEFINITIONS

What would you do with each of the following things whose names are printed in bold? Find and circle the *best* answer in each group of words. Use a dictionary if you need help.

1. What would you do with a **javelin**?
   - **a.** stamp it
   - **b.** roll it down a bowling alley
   - **c.** read it
   - **d.** throw it as far as you can

2. What would you do with a **jubilee**?
   - **a.** wax the floors with it
   - **b.** look at it under a magnifying glass
   - **c.** celebrate it with friends
   - **d.** spray it

3. What would you do with a **jackal**?
   - **a.** wear it on a cold night
   - **b.** photograph it at the zoo
   - **c.** use it to change a flat tire on a car
   - **d.** keep score with it

4. What would you do with **jade**?
   - **a.** bake it
   - **b.** dig with it
   - **c.** plant it
   - **d.** use it in a piece of jewelry

5. What would you do with a **jig**?
   - **a.** put it in a cage and feed it
   - **b.** bottle it
   - **c.** wash it
   - **d.** dance it

6. What would you do with a **junction**?
   - **a.** hold a committee meeting with it
   - **b.** stop at it and look for oncoming traffic
   - **c.** eat it with a spoon
   - **d.** auction it off

7. What would you do with **jute**?
   - **a.** pour it into a pitcher
   - **b.** chew it
   - **c.** blow into its holes to make music
   - **d.** make rope out of it

8. What would you do with **jasmine**?
   - **a.** plant it in a garden
   - **b.** roast it in the oven
   - **c.** write with it
   - **d.** sew it

9. What would you do with a **jackpot**?
   - **a.** cook jacks in it
   - **b.** bake beans in it
   - **c.** count the money you won from it
   - **d.** staple it

10. What would you do with a **joist**?
    - **a.** use it to support a ceiling
    - **b.** sing it in an opera
    - **c.** ride it in a race
    - **d.** tie it in a knot

Name ______________________________ Date ______________

What Would You Do with a Kaleidoscope?

VOCABULARY DEFINITIONS

What would you do with each of the following things whose names are printed in bold? Find and circle the *best* answer in each group of words. Use a dictionary if you need help.

1. What would you do with a **kaleidoscope**?
   - **a.** open a locked door with it
   - **b.** ride it to school
   - **c.** look through it to see different patterns
   - **d.** look through it to see microscopic germs

2. What would you do with a **kayak**?
   - **a.** paddle down the river in it
   - **b.** plant it in the ground
   - **c.** slice it and eat it for breakfast
   - **d.** sit under its shade on a hot day

3. What would you do with a **kimono**?
   - **a.** sing it
   - **b.** feed it at the petting zoo
   - **c.** dance to it
   - **d.** wear it

4. What would you do with a **knapsack**?
   - **a.** carry things in it when hiking
   - **b.** take a short nap on it
   - **c.** read it
   - **d.** use it to fix the plumbing

5. What would you do with a **knob**?
   - **a.** put medicine on it to reduce swelling
   - **b.** turn it to open a door
   - **c.** trim it
   - **d.** put notebook paper in it

6. What would you do with a **koala**?
   - **a.** drink it on a hot summer day
   - **b.** watch it hanging in a tree
   - **c.** cut its leaves
   - **d.** tie it in a knot

7. What would you do with a **knoll**?
   - **a.** sleep in it
   - **b.** use it to bake bread
   - **c.** hike up it while camping
   - **d.** cover it up

8. What would you do with a **kink**?
   - **a.** straighten it out
   - **b.** wear it to a party
   - **c.** listen to it chirp
   - **d.** paint it

9. What would you do with **kindling**?
   - **a.** tie a package with it
   - **b.** put a stamp on it
   - **c.** scrub it
   - **d.** use it to start a fire

10. What would you do with a **kettle**?
    - **a.** sew it
    - **b.** flip pancakes with it
    - **c.** boil water in it
    - **d.** wear it on your head for warmth

Name ______________________________ Date ______________

What Would You Do with a Legume?

VOCABULARY DEFINITIONS

What would you do with each of the following things whose names are printed in bold? Find and circle the *best* answer in each group of words. Use a dictionary if you need help.

1. What would you do with a **legume**?
   a. wrap it around your leg
   b. eat it
   c. paste it in a scrapbook
   d. curl your hair with it

2. What would you do with a **ladle**?
   a. saddle a horse with it
   b. use it to pour soup into bowls
   c. tie a bow with it
   d. use it to build a brick wall

3. What would you do with a **lever**?
   a. use it to lift something heavy
   b. cut meat with it
   c. pour coffee into it
   d. use it to sew a button on a shirt

4. What would you do with a **lilac**?
   a. see it in a pet store
   b. type a story on it
   c. get it X-rayed at the doctor's office
   d. plant it in a garden

5. What would you do with a **loom**?
   a. watch it jump around in its cage
   b. cook with it
   c. weave cloth on it
   d. jump rope with it

6. What would you do with **lotion**?
   a. rub it on your skin
   b. mix it with cream and drink it
   c. put it in a camera
   d. add it to a car's engine

7. What would you do with **lyrics**?
   a. play with them
   b. sing them in a school musical
   c. tape them to a wall
   d. wear them with suspenders

8. What would you do with **loot**?
   a. return it to its rightful owner
   b. buy it at a fish market
   c. fix your computer with it
   d. add it to cookie dough

9. What would you do with a **loft**?
   a. put it on the handlebars of a bike
   b. feed it to cattle
   c. use it as an art studio
   d. put bologna on it

10. What would you do with a **locket**?
    a. fasten a door with it
    b. tie a ponytail with it
    c. sew it on a sleeve
    d. wear it around your neck on a chain

Name ______________________________ Date ______________

What Would You Do with a Niche?

Vocabulary Definitions

What would you do with each of the following things whose names are printed in bold? Find and circle the *best* answer in each group of words. Use a dictionary if you need help.

1. What would you do with a **niche**?
   - **a.** fry it in a skillet
   - **b.** wear it to a prom
   - **c.** scratch it to stop the itching
   - **d.** put a vase or statue on display in it

2. What would you do with a **newt**?
   - **a.** polish it with wax
   - **b.** learn about it in biology class
   - **c.** wear it to block the sun
   - **d.** dig it up

3. What would you do with a **nectarine**?
   - **a.** play music on it
   - **b.** wear it scuba diving
   - **c.** eat it
   - **d.** wear it with a dress shirt

4. What would you do with a **newscast**?
   - **a.** watch it to learn what's happening in the world
   - **b.** put it on a broken arm
   - **c.** catch shrimp with it
   - **d.** repair a television set with it

5. What would you do with a **nominee**?
   - **a.** watch it cross the desert
   - **b.** vote for him or her
   - **c.** feed it fish at the aquarium
   - **d.** swallow it to reduce a fever

6. What would you do with a **nostril**?
   - **a.** smell and breathe through it
   - **b.** water flowers and grass with it
   - **c.** plant it during the spring
   - **d.** drill with it

7. What would you do with a **novelist**?
   - **a.** use it in the dark
   - **b.** make a list with it
   - **c.** get his or her autograph
   - **d.** watch him or her design a building

8. What would you do with a **nursery**?
   - **a.** take a person's temperature with it
   - **b.** operate on it
   - **c.** weigh it
   - **d.** buy young trees or plants at it

9. What would you do with a **numeral**?
   - **a.** add or subtract it in math
   - **b.** numb it to stop the pain
   - **c.** use it to write a letter
   - **d.** take it on vacation

10. What would you do with a **notation**?
    - **a.** move it to another location
    - **b.** put it in the refrigerator
    - **c.** read it to remind you of something
    - **d.** model it in a fashion show

Name ______________________ Date ____________

What Would You Do with a Route?

VOCABULARY DEFINITIONS

What would you do with each of the following things whose names are printed in bold? Find and circle the *best* answer in each group of words. Use a dictionary if you need help.

1. What would you do with a **route**?
   a. cheer the football team with it
   b. travel it to work each day
   c. pull it out of the ground
   d. dig a hole with it

2. What would you do with a **refrain**?
   a. sing it over and over in a song
   b. frame it and hang it on a wall
   c. reheat it and have it for dinner
   d. strain it

3. What would you do with **relish**?
   a. paint with it
   b. skate on it
   c. tell it to a best friend
   d. put it on a hot dog or hamburger

4. What would you do with a **repellant**?
   a. use it to tile a counter
   b. climb it
   c. put it on to keep mosquitoes away
   d. glue it

5. What would you do with a **ruffle**?
   a. eat it for dessert
   b. sew it on a dress
   c. float on the water with it
   d. attach it to the back of a boat

6. What would you do with a **reef**?
   a. hang it on the front door
   b. swim out to it
   c. eat it on a sandwich
   d. turn it upside down

7. What would you do with a **rookie**?
   a. teach him or her how to play
   b. bake it for a holiday
   c. watch it fly
   d. play it using a deck of cards

8. What would you do with a **rivet**?
   a. swim in it
   b. fasten metal beams together with it
   c. button it
   d. put a hot pot or pan on it

9. What would you do with **revenue**?
   a. drive on it
   b. put pizza in it to bake
   c. deposit it in a bank
   d. wash and rinse it thoroughly

10. What would you do with a **restriction**?
    a. obey it
    b. hum it
    c. restore it
    d. spend your summer vacation in it

Name ______________________ Date ____________

What Would You Do with a Toucan?

VOCABULARY DEFINITIONS

What would you do with each of the following things whose names are printed in bold? Find and circle the *best* answer in each group of words. Use a dictionary if you need help

1. What would you do with a **toucan**?
   - a. wrap it around your head
   - b. eat it for dessert
   - c. braid its long hair
   - d. watch it perched on a limb of a tree

2. What would you do with **tartar**?
   - a. tie it around your waist
   - b. pave a street with it
   - c. go to a dentist to have it removed
   - d. use it as a sauce for fish

3. What would you do with a **terminal**?
   - a. feed it to a monkey
   - b. wait for a bus or plane at it
   - c. listen to it in an orchestra
   - d. use it to dig a hole

4. What would you do with a **terrier**?
   - a. use it to make a blanket
   - b. ride it
   - c. use it to dry off after a shower
   - d. give it a dog biscuit

5. What would you do with a **throne**?
   - a. sit on it
   - b. play in it
   - c. skate on it in the winter
   - d. wear it on your head

6. What would you do with **tinsel**?
   - a. plug it in an electrical outlet
   - b. use it for decoration
   - c. freeze it
   - d. button it

7. What would you do with a **tempo**?
   - a. use him or her to replace an absent worker
   - b. play a musical instrument in time to it
   - c. go to it to worship and pray
   - d. write a letter with it

8. What would you do with **tongs**?
   - a. flip pancakes with them
   - b. speak into them to make your voice sound louder
   - c. use them to thread a needle for sewing
   - d. lift something with them

9. What would you do with **tinder**?
   - a. put it on meat to tenderize it
   - b. store things in it
   - c. use it to start a fire in a fireplace
   - d. use it to keep dry in the rain

10. What would you do with a **tiara**?
    - a. wear it on your head
    - b. take a ride in it
    - c. boil it
    - d. nail it into a wall

Name ______________________________ Date ______________

What Would You Do with a Vandal?

Vocabulary Definitions

What would you do with each of the following things whose names are printed in bold? Find and circle the *best* answer in each group of words. Use a dictionary if you need help.

1. What would you do with a **vandal**?
   - **a.** wear it on your foot at the beach
   - **b.** arrest him or her
   - **c.** pack things in it to take on a trip
   - **d.** play it with friends

2. What would you do with a **vacancy**?
   - **a.** put it in a playpen
   - **b.** find it on a menu at a restaurant
   - **c.** try to fill it
   - **d.** rinse it out

3. What would you do with a **vault**?
   - **a.** keep money or valuables in it
   - **b.** barbecue on it
   - **c.** sharpen a pencil with it
   - **d.** tell the direction of the wind with it

4. What would you do with **velvet**?
   - **a.** put it between two pieces of bread
   - **b.** hammer it
   - **c.** pour it over your head
   - **d.** use it to make a fancy dress or vest

5. What would you do with a **veranda**?
   - **a.** dance it to fast, upbeat music
   - **b.** play it in an orchestra
   - **c.** sit on it on a hot summer night
   - **d.** take it to a mechanic

6. What would you do with a **visa**?
   - **a.** wear it to block the sun from your face
   - **b.** blow it to call your dog inside
   - **c.** enjoy the view from it
   - **d.** show it to enter a foreign country

7. What would you do with a **vulture**?
   - **a.** photograph it at the zoo
   - **b.** turn it on to light up a room
   - **c.** pack it for your vacation
   - **d.** hit it over a net to an opposing team

8. What would you do with a **vocalist**?
   - **a.** listen to him or her sing a song
   - **b.** play it
   - **c.** spray your throat with it
   - **d.** recite it in class

9. What would you do with **vinegar**?
   - **a.** grow it on a vine
   - **b.** sharpen a knife with it
   - **c.** put it in your car's engine
   - **d.** flavor foods with it

10. What would you do with a **vessel**?
    - **a.** put it on under a shirt
    - **b.** hold something in it
    - **c.** eat dinner under it
    - **d.** put it on a calendar

Name ______________________________ Date ______________

Ladder Links #1

SPELLING AND VOCABULARY

The object of a ladder link puzzle is to change one word into another by changing only one letter at a time. You cannot change the order of the letters. Use the hints provided to determine the word. Here is an example of a ladder link:

Change FEAR into DEEP.

F E A R

<u>D</u> E A R (the F has been changed to D; the other letters remain the same)

D E <u>E</u> R (the A has been changed to E)

D E E <u>P</u> (the R has been changed to P)

1 Change HOLE into COOP.

H O L E

______________________ (to keep in a certain place or position)

______________________ (a covering for the head and neck)

______________________ (the metal rim of the basket in basketball)

C O O P

2 Change BEARS into FLAGS.

B E A R S

______________________ (feelings of being worried or excited)

______________________ (actions or deeds that show great courage or skill)

______________________ (musical notes or tones one half step below another)

F L A G S

Name ______________________ Date ____________

Ladder Links #2

SPELLING AND VOCABULARY

The object of a ladder link puzzle is to change one word into another by changing only one letter at a time. You cannot change the order of the letters. Use the clues to help you solve these ladder link puzzles.

1 Change FIVE into TAIL.

F I V E

______________ (a box to keep papers in order)

______________ (to put as much as possible into)

______________ (to drop to a lower place)

______________ (to not succeed)

T A I L

2 Change FOOT into HAND.

F O O T

______________ (what you eat)

______________ (loving and tender)

______________ (something that ties or binds)

______________ (a group of musicians playing together)

H A N D

Name ______________________________ Date ______________

Ladder Links #3

SPELLING AND VOCABULARY

The object of a ladder link puzzle is to change one word into another by changing only one letter at a time. You cannot change the order of the letters. Use the clues to help you solve these ladder link puzzles.

1 Change FIRE into HEAT.

F I R E

______________ (to select someone for a job)

______________ (at or in this place; not there)

______________ (group of cattle)

______________ (where your brain is located)

H E A T

2 Change STILE into TEARS.

S T I L E

______________ (not fresh)

______________ (a long, steady look)

______________ (heavenly bodies that shine by their own light)

______________ (burns or scorches the surface of)

T E A R S

Name ______________________ Date ____________

Ladder Links #4

SPELLING AND VOCABULARY

The object of a ladder link puzzle is to change one word into another by changing only one letter at a time. You cannot change the order of the letters. Use the clues to help you solve these ladder link puzzles.

1 Change SMART into BLACK.

S M A R T

______________ (to begin)

______________ (bare; plain)

______________ (a large, neat pile of something)

______________ (not tight; loose)

B L A C K

2 Change HOOK into FISH.

H O O K

______________ (to keep your eyes fixed on)

______________ (to rob or take by force)

______________ (missing or misplaced)

______________ (a series of names written down one after another)

______________ (a hand with the fingers closed tightly)

F I S H

Name ______________________________ Date ______________

Ladder Links #5

SPELLING AND VOCABULARY

The object of a ladder link puzzle is to change one word into another by changing only one letter at a time. You cannot change the order of the letters. Use the clues to help you solve these ladder link puzzles.

1 Change HOPE into GRAD.

H O P E

______________ (an opening in or through something; a break or tear)

______________ (to keep in a certain position or place)

______________ (a heavy, yellow metal used in jewelry)

______________ (to excite or rouse)

G R A D

2 Change SLOW into PENT.

S L O W

______________ (a narrow, straight opening)

______________ (a thin, narrow strip of wood)

______________ (a chair or bench to sit on)

______________ (the past tense of *send*)

P E N T

Name ______________________________ Date ______________

Ladder Links #6

SPELLING AND VOCABULARY

The object of a ladder link puzzle is to change one word into another by changing only one letter at a time. You cannot change the order of the letters. Use the clues to help you solve these ladder link puzzles.

1 Change NEVER into TIMES.

N E V E R

______________ (a bar that can be moved to turn something)

______________ (an organ in the human body that stores bile)

______________ (a cat has nine of these)

______________ (lemon-like fruits with greenish-yellow skin)

T I M E S

2 Change HARD into EASE.

H A R D

______________ (a division of a hospital)

______________ (a magician's stick)

______________ (to become less in size or strength)

______________ (a stick that a person may carry for help in walking)

______________ (a container for holding and protecting something)

E A S E

Name ______________________________ Date ______________

Ladder Links #7

SPELLING AND VOCABULARY

The object of a ladder link puzzle is to change one word into another by changing only one letter at a time. You cannot change the order of the letters. Use the clues to help you solve these ladder link puzzles.

1 Change CHEW into SLED.

C H E W

______________________ (food)

______________________ (allow to be seen; reveal)

______________________ (not fast or quick in moving or working)

______________________ (past tense of *slay*)

S L E D

2 Change BLANK into GRAIN.

B L A N K

______________________ (mild and plain in flavoring)

______________________ (trademark)

______________________ (to weave three or more strands together)

______________________ (the center of thought, memory, and feelings)

G R A I N

Name ______________________________ Date ______________

Ladder Links #8

SPELLING AND VOCABULARY

The object of a ladder link puzzle is to change one word into another by changing only one letter at a time. You cannot change the order of the letters. Use the clues to help you solve these ladder link puzzles.

1 Change BEAT into MANE.

B E A T

______________ (the flesh of an animal that is used as food)

______________ (to change from a solid to a liquid by heat)

______________ (barley or grain used in making drinks)

______________ (of or for men or boys)

M A N E

2 Change WARN into COLD.

W A R N

______________ (not cool but not hot; a pleasant heat)

______________ (to turn aside or keep off)

______________ (paper, often with printing on it)

______________ (thin rope or a thick string)

C O L D

Name ______________________ Date ____________

Ladder Links #9

SPELLING AND VOCABULARY

The object of a ladder link puzzle is to change one word into another by changing only one letter at a time. You cannot change the order of the letters. Use the clues to help you solve these ladder link puzzles.

1 Change TOOL into BALE.

T O O L

______________ (to work hard; to labor)

______________ (thin metal sheet)

______________ (not succeed)

______________ (to tumble or collapse)

______________ (any round object; sphere)

B A L E

2 Change SAGE into WELL.

S A G E

______________ (one side of a leaf of paper in a book or magazine)

______________ (not bright; dim)

______________ (to become tired of something)

______________ (a structure that forms the side of a building)

W E L L

Name ______________________________ Date ______________

Ladder Links #10

SPELLING AND VOCABULARY

The object of a ladder link puzzle is to change one word into another by changing only one letter at a time. You cannot change the order of the letters. Use the clues to help you solve these ladder link puzzles.

1 Change COAL into FINE.

C O A L

______________________ (a very young horse)

______________________ (a person who lacks good sense; a silly person)

______________________ (what you eat)

______________________ (loving and tender)

______________________ (to discover)

F I N E

2 Change WHALE into SPOUT.

W H A L E

______________________ (a rock formed of hardened clay)

______________________ (to have or use together)

______________________ (land at the edge of a sea or lake)

______________________ (a tiny cell produced by certain plants and animals)

______________________ (a game done for exercise or pleasure)

S P O U T

Name ______________________________ Date ______________

Noun Analogies

ANALOGIES

An **analogy** is the relationship between one pair of words that serves as the basis for the creation of another pair of words. Find the word that *best* completes each analogy. Circle the letter of your choice.

1 snake : reptile :: beetle :
- **a.** wasp
- **b.** insect
- **c.** fly
- **d.** lizard

2 finger : hand :: toe :
- **a.** shoe
- **b.** nail
- **c.** leg
- **d.** foot

3 camel : desert :: monkey :
- **a.** tail
- **b.** ocean
- **c.** jungle
- **d.** banana

4 author : writer :: guide :
- **a.** leader
- **b.** safari
- **c.** vacation
- **d.** trip

5 tulip : flower :: elm :
- **a.** oak
- **b.** tree
- **c.** flower
- **d.** leaf

6 minute : hour :: day :
- **a.** pound
- **b.** second
- **c.** week
- **d.** ounce

7 ear : sound :: eye :
- **a.** glass
- **b.** smell
- **c.** taste
- **d.** sight

8 canary : bird :: tuna :
- **a.** sandwich
- **b.** salmon
- **c.** fish
- **d.** reptile

9 bow : arrow :: salt :
- **a.** spice
- **b.** pepper
- **c.** tomato
- **d.** barbecue

10 knee : leg :: elbow :
- **a.** hand
- **b.** ankle
- **c.** wrist
- **d.** arm

Name ______________________________ Date ______________

Verb Analogies

ANALOGIES

An **analogy** is the relationship between one pair of words that serves as the basis for the creation of another pair of words. Find the word that *best* completes each analogy. Circle the letter of your choice.

1. climb : ascend :: drop :
   a. pull
   b. fall
   c. drag
   d. drain

2. arrive : depart :: give :
   a. extend
   b. offer
   c. live
   d. take

3. singe : burn :: search :
   a. perch
   b. glitter
   c. seek
   d. adjust

4. vanish : appear :: fail :
   a. flunk
   b. flail
   c. pale
   d. pass

5. listen : talk :: remember :
   a. forget
   b. recall
   c. think
   d. regret

6. twist : bend :: repair :
   a. offer
   b. break
   c. destroy
   d. mend

7. add : subtract :: stop :
   a. halt
   b. start
   c. step
   d. end

8. shake : quiver :: beat :
   a. beet
   b. soothe
   c. pound
   d. push

9. help : hinder :: buy :
   a. purchase
   b. bye
   c. sell
   d. return

10. protect : guard :: fool :
   a. begin
   b. drool
   c. march
   d. trick

Name ______________________________ Date ______________

Adjective Analogies

ANALOGIES

An **analogy** is the relationship between one pair of words that serves as the basis for the creation of another pair of words. Find the word that *best* completes each analogy. Circle the letter of your choice.

1. large : huge :: tiny :
   - **a.** dull
   - **b.** enormous
   - **c.** big
   - **d.** small

2. many : few :: guilty :
   - **a.** innocent
   - **b.** gaudy
   - **c.** graceful
   - **d.** jealous

3. far : near :: shallow :
   - **a.** watery
   - **b.** deep
   - **c.** thin
   - **d.** shaky

4. moist : damp :: dry :
   - **a.** desert
   - **b.** wet
   - **c.** sunny
   - **d.** parched

5. dangerous : safe :: serious :
   - **a.** somber
   - **b.** silly
   - **c.** mean
   - **d.** idle

6. soft : hard :: dull :
   - **a.** boring
   - **b.** bashful
   - **c.** lonesome
   - **d.** sharp

7. dark : murky :: concealed :
   - **a.** thick
   - **b.** hidden
   - **c.** weak
   - **d.** showy

8. dirty : clean :: calm :
   - **a.** upset
   - **b.** peaceful
   - **c.** tranquil
   - **d.** pretty

9. timid : shy :: sick :
   - **a.** healthy
   - **b.** ill
   - **c.** well
   - **d.** athletic

10. solid : flimsy :: fancy :
    - **a.** rough
    - **b.** dressy
    - **c.** ornate
    - **d.** plain

Name ______________________________ Date ______________

Analogy Challenge #1

ANALOGIES

An **analogy** is the relationship between one pair of words that serves as the basis for the creation of another pair of words. Find the word that *best* completes each analogy. Circle the letter of your choice.

1. flour : flower :: veil :
   a. bride
   b. vale
   c. lace
   d. rein

2. liquid : milk :: gas :
   a. air
   b. solid
   c. automobile
   d. pump

3. young : old :: absent :
   a. tardy
   b. late
   c. present
   d. sick

4. sandal : foot :: hat :
   a. glove
   b. cap
   c. helmet
   d. head

5. maps : atlas :: words :
   a. speak
   b. dictionary
   c. talk
   d. alphabet

6. man : father :: woman :
   a. sister
   b. grandmother
   c. baby
   d. mother

7. shell : turtle :: crust :
   a. food
   b. rust
   c. pie
   d. dough

8. glimpse : see :: make :
   a. attack
   b. destroy
   c. view
   d. create

9. tree : forest :: flower :
   a. petal
   b. garden
   c. daisy
   d. grass

10. fur : bear :: scales :
    a. fish
    b. bird
    c. feather
    d. tiger

Name ______________________________ Date ______________

Analogy Challenge #2

ANALOGIES

An **analogy** is the relationship between one pair of words that serves as the basis for the creation of another pair of words. Find the word that *best* completes each analogy. Circle the letter of your choice.

1. student : class :: player :
   - **a.** football
   - **b.** sport
   - **c.** goalie
   - **d.** team

2. inch : foot :: month :
   - **a.** yard
   - **b.** pound
   - **c.** year
   - **d.** second

3. racket : tennis :: bat :
   - **a.** football
   - **b.** round
   - **c.** mammal
   - **d.** baseball

4. thermometer : temperature :: clock :
   - **a.** time
   - **b.** watch
   - **c.** chime
   - **d.** scale

5. airplane : pilot :: bus :
   - **a.** wheels
   - **b.** driver
   - **c.** passenger
   - **d.** taxi

6. isle : aisle :: threw :
   - **a.** ball
   - **b.** tossed
   - **c.** through
   - **d.** pitched

7. earth : planet :: sun :
   - **a.** hot
   - **b.** yellow
   - **c.** solar
   - **d.** star

8. saffron : yellow :: crimson :
   - **a.** green
   - **b.** red
   - **c.** blue
   - **d.** purple

9. baton : conductor :: scalpel :
   - **a.** surgeon
   - **b.** knife
   - **c.** sharp
   - **d.** sculptor

10. broccoli : vegetable :: orange :
    - **a.** lemon
    - **b.** rind
    - **c.** fruit
    - **d.** dairy

Name ______________________________ Date ______________

Analogy Challenge #3

ANALOGIES

An **analogy** is the relationship between one pair of words that serves as the basis for the creation of another pair of words. Find the word that *best* completes each analogy. Circle the letter of your choice.

1. blame : criticize :: compliment :
   - **a.** praise
   - **b.** scold
   - **c.** argue
   - **d.** match

2. ice : melt :: candle :
   - **a.** wax
   - **b.** light
   - **c.** hot
   - **d.** burn

3. fly : airplane :: sail :
   - **a.** sale
   - **b.** boat
   - **c.** fish
   - **d.** canvas

4. tusk : walrus :: tooth :
   - **a.** teeth
   - **b.** enamel
   - **c.** lion
   - **d.** cavity

5. pinto : bean :: yam :
   - **a.** ham
   - **b.** tuber
   - **c.** russet
   - **d.** fruit

6. child : children :: mouse :
   - **a.** mammal
   - **b.** mouses
   - **c.** mice
   - **d.** meese

7. finger : snap :: eye :
   - **a.** lash
   - **b.** wink
   - **c.** lid
   - **d.** pupil

8. muscle : mussel :: straight :
   - **a.** crooked
   - **b.** narrow
   - **c.** strait
   - **d.** aligned

9. idle : busy :: expand :
   - **a.** contract
   - **b.** enlarge
   - **c.** extend
   - **d.** erase

10. strum : guitar :: blow :
    - **a.** drum
    - **b.** trumpet
    - **c.** piano
    - **d.** harp

Name ______________________________ Date ______________

Analogy Challenge #4

ANALOGIES

An **analogy** is the relationship between one pair of words that serves as the basis for the creation of another pair of words. Find the word that *best* completes each analogy. Circle the letter of your choice.

1. emerald : green :: beige :
   a. color
   b. black
   c. brown
   d. orange

2. load : lode :: weight :
   a. measurement
   b. pound
   c. ton
   d. wait

3. march : parade :: sing :
   a. sign
   b. chorus
   c. band
   d. music

4. glove : hand :: boot :
   a. shoe
   b. arm
   c. sock
   d. foot

5. knight : night :: chews :
   a. eats
   b. munches
   c. choose
   d. chose

6. hive : bees :: colony :
   a. birds
   b. tigers
   c. moths
   d. ants

7. kangaroo : joey :: giraffe :
   a. calf
   b. tall
   c. colt
   d. kit

8. geese : goose :: feet :
   a. cleat
   b. foot
   c. yard
   d. s area

9. arctic : cold :: tropic :
   a. jungle
   b. lizard
   c. hot
   d. snow

10. nose : head :: finger :
    a. hand
    b. polish
    c. nail
    d. thigh

Name ______________________________ Date ______________

Synonym and Antonym Analogies

ANALOGIES

An **analogy** is the relationship between one pair of words that serves as the basis for the creation of another pair of words. Find the word that *best* completes each analogy. Circle the letter of your choice.

1. off : on :: always :
   - **a.** about
   - **b.** already
   - **c.** before
   - **d.** never

2. stale : fresh :: rare :
   - **a.** common
   - **b.** steak
   - **c.** wealthy
   - **d.** tender

3. start : begin :: trick :
   - **a.** magician
   - **b.** fool
   - **c.** treat
   - **d.** trip

4. occur : happen :: risk :
   - **a.** wrist
   - **b.** rink
   - **c.** danger
   - **d.** storm

5. friend : foe :: rude :
   - **a.** crude
   - **b.** polite
   - **c.** nasty
   - **d.** mean

6. permanent : temporary :: omit :
   - **a.** reject
   - **b.** include
   - **c.** admit
   - **d.** dismiss

7. awkward : clumsy :: dense :
   - **a.** thick
   - **b.** thin
   - **c.** cents
   - **d.** narrow

8. backward : forward :: follow :
   - **a.** fellow
   - **b.** copy
   - **c.** lead
   - **d.** fear

9. sink : float :: advance :
   - **a.** retreat
   - **b.** think
   - **c.** plan
   - **d.** imitate

10. display : exhibit :: caution :
    - **a.** clean
    - **b.** tremble
    - **c.** repair
    - **d.** warn

Name ______________________ Date ____________

On the Menu Analogies

ANALOGIES

An **analogy** is the relationship between one pair of words that serves as the basis for the creation of another pair of words. Find the word that *best* completes each analogy. Circle the letter of your choice.

1 peel : banana :: shell :
- **a.** egg
- **b.** oval
- **c.** skin
- **d.** fruit

2 cabbage : vegetable :: apple :
- **a.** pear
- **b.** fruit
- **c.** pie
- **d.** tart

3 knead : dough :: scramble :
- **a.** fish
- **b.** meat
- **c.** bread
- **d.** egg

4 candy : sweet :: lemon :
- **a.** fruit
- **b.** yellow
- **c.** sour
- **d.** lime

5 meatballs : spaghetti :: peanut butter :
- **a.** nut
- **b.** jelly
- **c.** cream
- **d.** pudding

6 paprika : spice :: milk :
- **a.** cow
- **b.** tea
- **c.** coffee
- **d.** beverage

7 ketchup : tomato :: cheese :
- **a.** milk
- **b.** cheddar
- **c.** please
- **d.** meat

8 teaspoon : tablespoon :: pint :
- **a.** ton
- **b.** inch
- **c.** quart
- **d.** pound

9 trout : fish :: pork :
- **a.** meat
- **b.** meet
- **c.** barbecue
- **d.** fries

10 steak : stake :: cereal :
- **a.** breakfast
- **b.** serial
- **c.** oats
- **d.** banana

Name ______________________ Date ____________

Sports Analogies

ANALOGIES

An **analogy** is the relationship between one pair of words that serves as the basis for the creation of another pair of words. Find the word that *best* completes each analogy. Circle the letter of your choice.

1. goal : hockey :: touchdown :
   - **a.** basketball
   - **b.** golf
   - **c.** soccer
   - **d.** football

2. bowl : alley :: golf :
   - **a.** gulf
   - **b.** course
   - **c.** tee
   - **d.** club

3. team : teem :: surf :
   - **a.** water
   - **b.** ocean
   - **c.** tide
   - **d.** serf

4. tennis : sport :: checkers :
   - **a.** king
   - **b.** board
   - **c.** game
   - **d.** jump

5. volleyball : sport :: rose :
   - **a.** rows
   - **b.** thorn
   - **c.** petal
   - **d.** flower

6. dribble : basketball :: punt :
   - **a.** football
   - **b.** puck
   - **c.** stick
   - **d.** hoop

7. ring : boxing :: court :
   - **a.** lawyer
   - **b.** tennis
   - **c.** wrestling
   - **d.** bowling

8. course : coarse :: pedal :
   - **a.** bicycle
   - **b.** race
   - **c.** peddle
   - **d.** medal

9. victory : defeat :: offense :
   - **a.** precise
   - **b.** exempt
   - **c.** definite
   - **d.** defense

10. coach : instructor :: thief :
    - **a.** robber
    - **b.** police
    - **c.** bank
    - **d.** prisoner

Name ______________________________ Date ______________

Adventure Analogies

ANALOGIES

An **analogy** is the relationship between one pair of words that serves as the basis for the creation of another pair of words. Find the word that *best* completes each analogy. Circle the letter of your choice.

1. north : south :: east :
   - **a.** travel
   - **b.** west
   - **c.** map
   - **d.** right

2. journey : trip :: trail :
   - **a.** path
   - **b.** trip
   - **c.** mountain
   - **d.** road

3. canteen : water :: vase :
   - **a.** pot
   - **b.** urn
   - **c.** jar
   - **d.** flowers

4. air : glider :: river :
   - **a.** stream
   - **b.** canoe
   - **c.** ocean
   - **d.** airplane

5. freight : cargo :: course :
   - **a.** coarse
   - **b.** hoarse
   - **c.** route
   - **d.** rough

6. lion : roar :: bird :
   - **a.** tree
   - **b.** chirp
   - **c.** mammal
   - **d.** feather

7. warn : caution :: shrink :
   - **a.** shrivel
   - **b.** enlarge
   - **c.** shark
   - **d.** grow

8. solo : one :: duet :
   - **a.** diet
   - **b.** four
   - **c.** two
   - **d.** sing

9. leaf : tree :: leg :
   - **a.** pencil
   - **b.** neck
   - **c.** fish
   - **d.** chair

10. wood : would :: jamb :
    - **a.** lamb
    - **b.** jelly
    - **c.** jam
    - **d.** stuck

Name ______________________________ Date ______________

People Analogies

ANALOGIES

An **analogy** is the relationship between one pair of words that serves as the basis for the creation of another pair of words. Find the word that *best* completes each analogy. Circle the letter of your choice.

1. actor : actress :: hero :
   - **a.** sandwich
   - **b.** heroine
   - **c.** male
   - **d.** female

2. principal : principle :: prophet :
   - **a.** profit
   - **b.** priest
   - **c.** rabbi
   - **d.** money

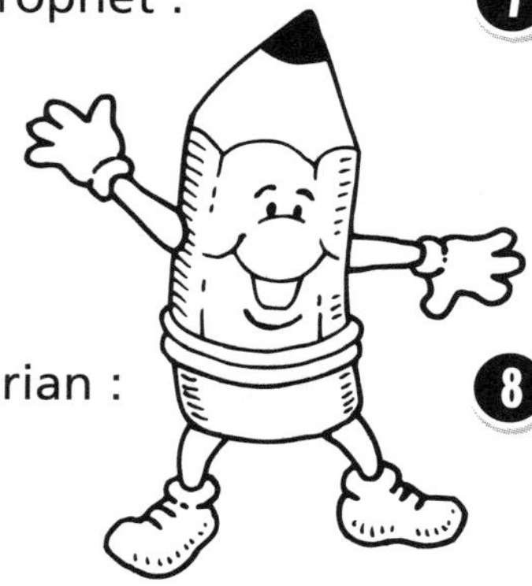

3. doctor : hospital :: librarian :
   - **a.** books
   - **b.** silence
   - **c.** library
   - **d.** restaurant

4. umpire : baseball :: referee :
   - **a.** basketball
   - **b.** whistle
   - **c.** judge
   - **d.** player

5. artist : easel :: plumber :
   - **a.** electrician
   - **b.** wrench
   - **c.** scissors
   - **d.** ocean

6. scientist : laboratory :: secretary :
   - **a.** telephone
   - **b.** computer
   - **c.** office
   - **d.** boss

7. maid : made :: baron :
   - **a.** ambassador
   - **b.** barren
   - **c.** baton
   - **d.** chief

8. judge : courtroom :: chef :
   - **a.** oven
   - **b.** cook
   - **c.** apron
   - **d.** kitchen

9. aunt : uncle :: niece :
   - **a.** nice
   - **b.** nephew
   - **c.** sister
   - **d.** mother

10. king : queen :: prince :
    - **a.** kingdom
    - **b.** mayor
    - **c.** governor
    - **d.** princess

Name ______________________________ Date ____________

Palindrome Fun

PALINDROMES, SPELLING, VOCABULARY

A **palindrome** is a word that reads the same forward and backward.

Read each clue and then complete the palindrome by filling in the missing consonants.

1	father	___ a ___
2	a joke	___ a ___
3	a device used by police to measure speed	___ a ___ a ___
4	a fast vehicle (2 words)	___ a ___ e ___ a ___
5	night before a holiday	e ___ e
6	flat and even	___ e ___ e ___
7	female sheep	e ___ e
8	look	___ e e ___
9	mention, call attention to	___ e ___ e ___
10	of a city	___ i ___ i ___
11	mother	___ o ___
12	twelve o'clock in the daytime	___ o o ___
13	short blast from a whistle or horn	___ o o ___
14	part of a motor that turns or rotates	___ o ___ o ___
15	songs sung by only one person	___ o ___ o ___

Name ________________________________ Date ______________

Homophones

HOMOPHONES, SPELLING, VOCABULARY

Homophones are words that sound alike but have different spellings and meanings such as *ant* and *aunt*.

Unscramble the words in Column A, and write the words on the lines on the right. Then match them to their homophones in Column B. Write the letter on the line. The first one has been done for you.

Column A			Column B
J	1 usm	sum	A. whole
____	2 ese	____________	B. pale
____	3 nus	____________	C. knows
____	4 rbae	____________	D. stare
____	5 eder	____________	E. dear
____	6 frea	____________	F. sea
____	7 lipa	____________	G. fair
____	8 ncet	____________	H. tale
____	9 soen	____________	I. son
____	10 helo	____________	J. some
____	11 sarit	____________	K. bare
____	12 lait	____________	L. sent

Name ______________________________ Date ______________

Hidden Animals

SPELLING

The name of an animal is hidden in two or more adjacent words in each sentence. Find and write the name of each animal. Use the words in the box if you need help. The first one has been done for you.

bat	bear	ewe	dog
goat	horse	leopard	ostrich
hen	gerbil	mouse	toad

1. Find a place to sit on the benc**h or se**at by the pool. ____horse____
2. I expect her to do great things with her life after she graduates. ______________
3. The cub at the zoo was the focus of attention. ______________
4. Jason wants to be a radiologist and work in a busy hospital. ______________
5. "Mom, Elmo used all of my notebook paper," grumbled Zack. ______________
6. "Do you want a hamburger, Bill, or would you prefer a hot dog?" he asked. ______________
7. She said, "Leo, pardon me for interrupting, but you have an important phone call." ______________
8. You can see we have a lot of experience when it comes to setting up a campsite. ______________
9. I don't understand why he never gets anywhere on time; he's always late. ______________
10. If you plan to go, a taxi can meet you at your hotel and drive you to the airport. ______________
11. I have to add these three columns of numbers to get the sum. ______________
12. Archaeologists have discovered many lost riches in their excavations. ______________

Name ______________________ Date ____________

Hidden Foods

SPELLING

The name of a food is hidden in two or more adjacent words in each sentence below. Find and write the name of each food. Use the words in the box if you need help. The first one has been done for you.

apple	mango	bean	pear
beef	tofu	lime	toast
fish	tomato	peach	turnip

1. This delicious chicken sou**p ear**ns you high praise as a chef. pear
2. One man goes to work in the city, but another works on a farm. ____________
3. When it is my turn, I plan to hit a home run for my team. ____________
4. The president of the organization asked our company to fund their next charity ball. ____________
5. Please say hello to Mat on Thursday when he comes to visit. ____________
6. The plant was dying, so the gardener had to clip each leaf that was dried out. ____________
7. When I surf, I share my board with a friend whose surfboard was stolen. ____________
8. After I graduate from college, I plan to be an architect and design skyscrapers. ____________
9. The bee flew around our picnic table but didn't bother us. ____________
10. The dog was slim enough to slip through the hole in the fence. ____________
11. If you decide to take a nap, please turn the television off so we don't waste electricity. ____________
12. When I go to a stadium for a football game, I bring along my camera so I can get some good shots of the team. ____________

Name ______________________________ Date ______________

Alphabetical Bubbles

ALPHABETICAL ORDER, SPELLING, SYLLABLES

In each bubble, write words that contain letters that are in alphabetical order such as *art* and *dim*. Notice that letters can be skipped in between, but each letter that appears in the word must be in alphabetical order.

One-Syllable Words

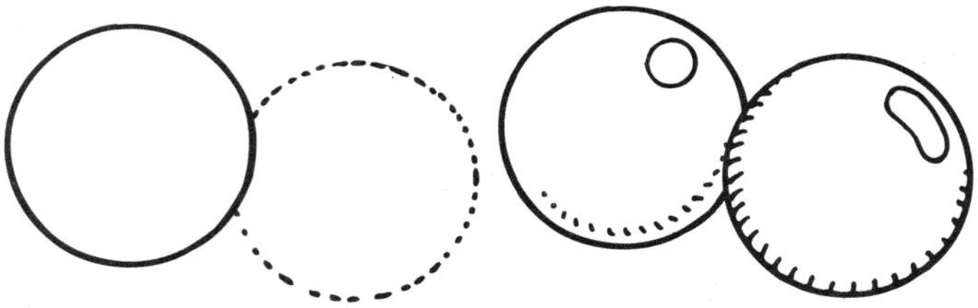

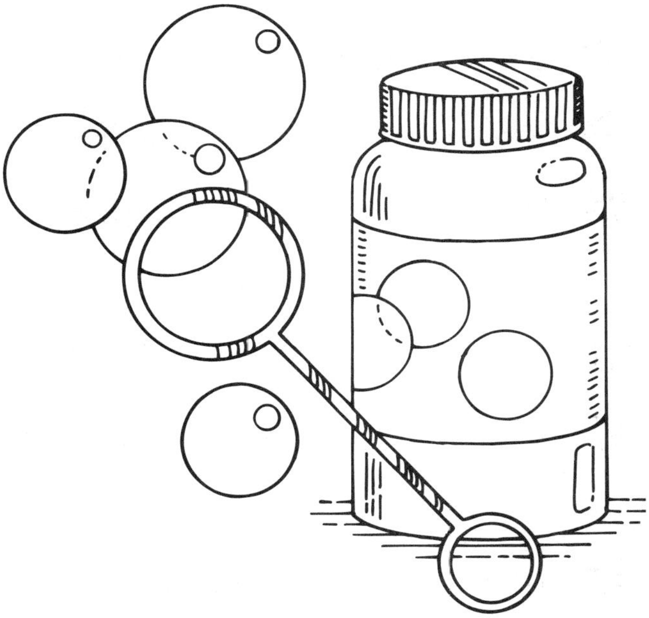

Two-Syllable Words

Three-Letter Words

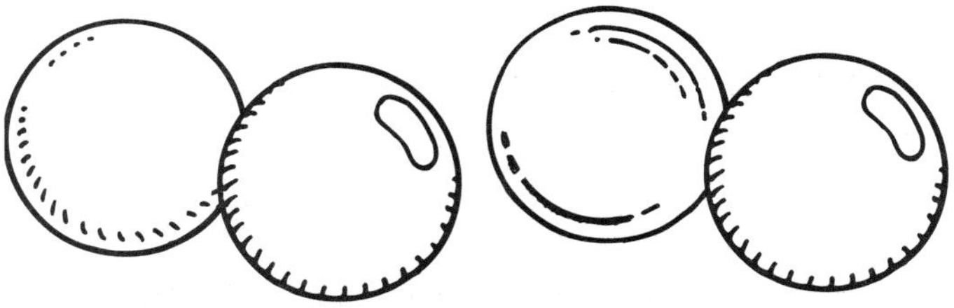

Sentences that contain six words with the words in alphabetical order

Just for Fun

Write the longest sentence you can (that makes sense) with each word in alphabetical order.

Name ______________________________ Date ______________

Hidden Trios

Spelling and Vocabulary

Each of these three-letter words can be found in other, longer words. The three letters may appear in the front, middle, or end of the longer word. For example, *her* is found in ***her**e*, *t**her**e,* and *ot**her***. List at least three longer words for each "hidden trio."

are

____

____

____

____

____

hip

____

____

____

____

____

sit

____

____

____

____

____

top

____

____

____

____

____

way

____

____

____

____

____

ten

____

____

____

____

____

Name ______________________________ Date ______________

More Hidden Trios

Spelling and Vocabulary

Each of these three-letter words can be found in other, longer words. The three letters may appear in the front, middle, or end of the longer word. For example, *her* is found in ***her**e*, *t**here**,* and *ot**her***. List at least three longer words for each "hidden trio."

lot

____

____

____

____

____

sea

____

____

____

____

____

own

____

____

____

____

____

row

____

____

____

____

____

let

____

____

____

____

____

set

____

____

____

____

____

Name ______________________________ Date ______________

Four-Letter Riddle Words

SPELLING AND VOCABULARY

In each row, write the four-letter word that best answers the clue. Use the words in the box if you need help. Then, going in order, write the letters in the shaded boxes on the lines below to solve the riddle. The first one has been done for you.

When is a piece of chicken like a musical instrument?

1. air that is moving
2. firm; not soft
3. crooked and curved
4. to bring together; connect
5. a long walk in the country or woods
6. to stop doing something
7. to not succeed; to not do what one tried to do
8. free from harm or danger
9. boast
10. a plan or purpose
11. not completely cooked; partly raw
12. a list of foods offered in a restaurant
13. unkind, rude, or nasty
14. put in danger
15. to put into words; to say
16. the front part of the lower jaw
17. to bite or grind up with the teeth
18. to build, create, or put together

1.	W	I	N	D
2.				
3.				
4.				
5.				
6.				
7.				
8.				
9.				
10.				
11.				
12.				
13.				
14.				
15.				
16.				
17.				
18.				

bent	chin	hard	join	mean	rare
brag	fail	hike	make	menu	risk
chew	wind	idea	tell	quit	safe

Name ______________________________ Date ______________

Odd Word Out

Classifying and Vocabulary

In each row, find the word that does not belong. Circle the word and explain why it does not belong with the other words. Use a dictionary if you need help.

1	lavender	paint	brown	gray
2	index	glossary	preface	encyclopedia
3	chef	bake	broil	boil
4	inning	homer	huddle	bunt
5	violet	petunia	sycamore	carnation
6	curse	modem	menu	cursor
7	gold	nickel	silver	dollar
8	nose	mouth	foot	eye
9	alligator	cobra	boa	python
10	dribble	hoop	rebound	goalie

Name ______________________ Date __________

More Odd Word Out

CLASSIFYING AND VOCABULARY

In each row, find the word that does not belong. Circle the word and explain why it does not belong with the other words. Use a dictionary if you need help.

1	around	through	many	between
2	Boston	Memphis	Detroit	Ohio
3	alas	oh	ouch	atlas
4	violin	harp	cello	tango
5	Australia	Italy	Denmark	Greece
6	yard	week	month	year
7	circle	square	tripod	triangle
8	Asia	Africa	Europe	London
9	route	petal	leaf	root
10	hurricane	tornado	cyclone	motorcycle

Name ______________________________ Date ______________

Wacky Riddles

VOCABULARY

Can you solve each riddle? Write the answers on the lines.

1. I'm full of holes but can still hold a lot of water. What am I?

   ____

2. I'm placed on a table, cut, but never eaten for a meal. What am I?

   ____

3. I always get wet when I'm drying. What am I?

   ____

4. What belongs to you, but your family and friends use it more than you do?

   ____

5. I always get larger the more you take away from me. What am I?

   ____

6. What can you always keep even after you give it to another person?

   ____

7. The more of these you take, the more you leave behind. What are they?

   ____

8. I have been around for millions of years, but I'm no more than a month old. What am I?

   ____

Name ______________________________ Date ______________

Verb Tense Bull's-Eye

VERB TENSES AND SPELLING

Starting at number 1 and moving clockwise, fill in the present tense form of each past participle verb. Place one letter in each box. The last letter of each word will begin the next word in the puzzle.

Clues
1. eaten
2. thrown
3. worn
4. rung
5. grown
6. written

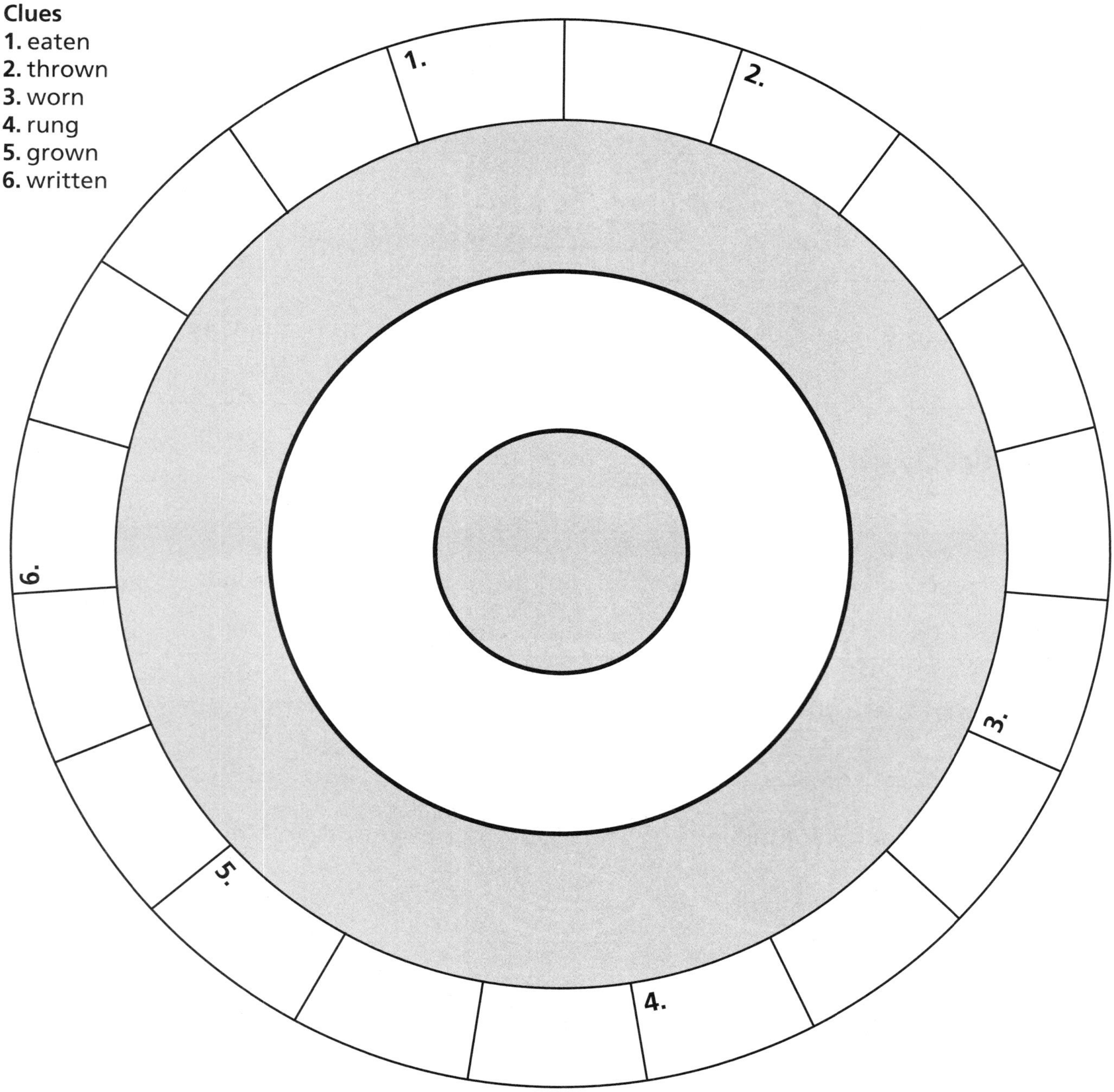

Just for Fun
Make up a verb tense bull's-eye puzzle for a classmate to solve.

Name ______________________________ Date ______________

Synonym Bull's-Eye

SYNONYMS AND SPELLING

Starting at number 1 and moving clockwise, fill in a synonym using the clues provided. Place one letter in each box. The last letter of each word will begin the next word in the puzzle.

Clues

1. moist
2. pledge
3. whole
4. forgive
5. finish
6. choose
7. deserve
8. observed

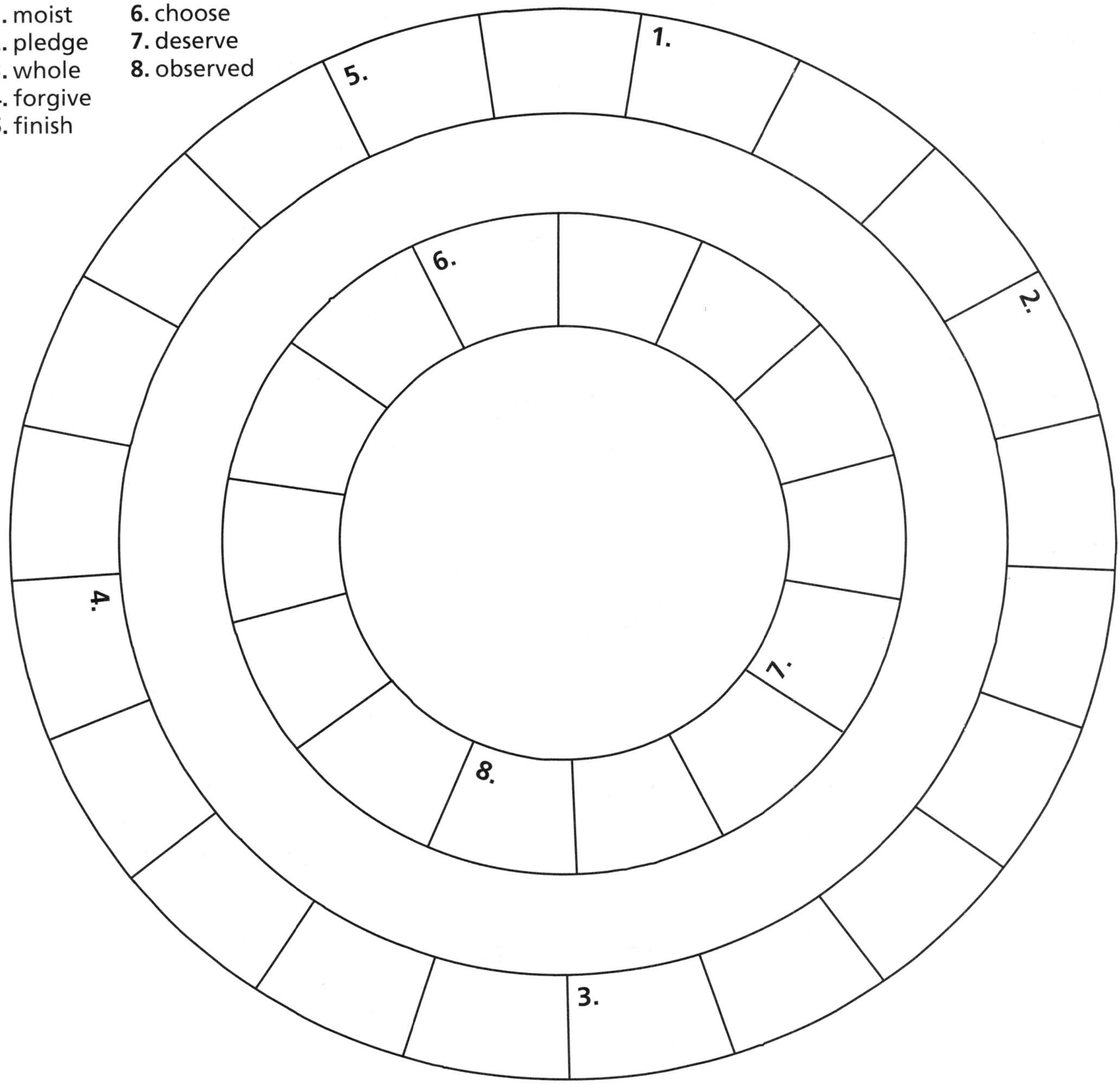

Just for Fun

Make up a synonym bull's-eye puzzle for a classmate to solve.

Name ______________________________ Date ______________

Antonym Bull's-Eye

ANTONYMS AND SPELLING

Starting at number 1 and moving clockwise, fill in an antonym using the clues provided. Place one letter in each box. The last letter of each word will begin the next word in the puzzle.

Clues

1. crooked
2. wild
3. vague
4. false
5. full
6. no
7. go
8. push
9. win
10. good
11. hates

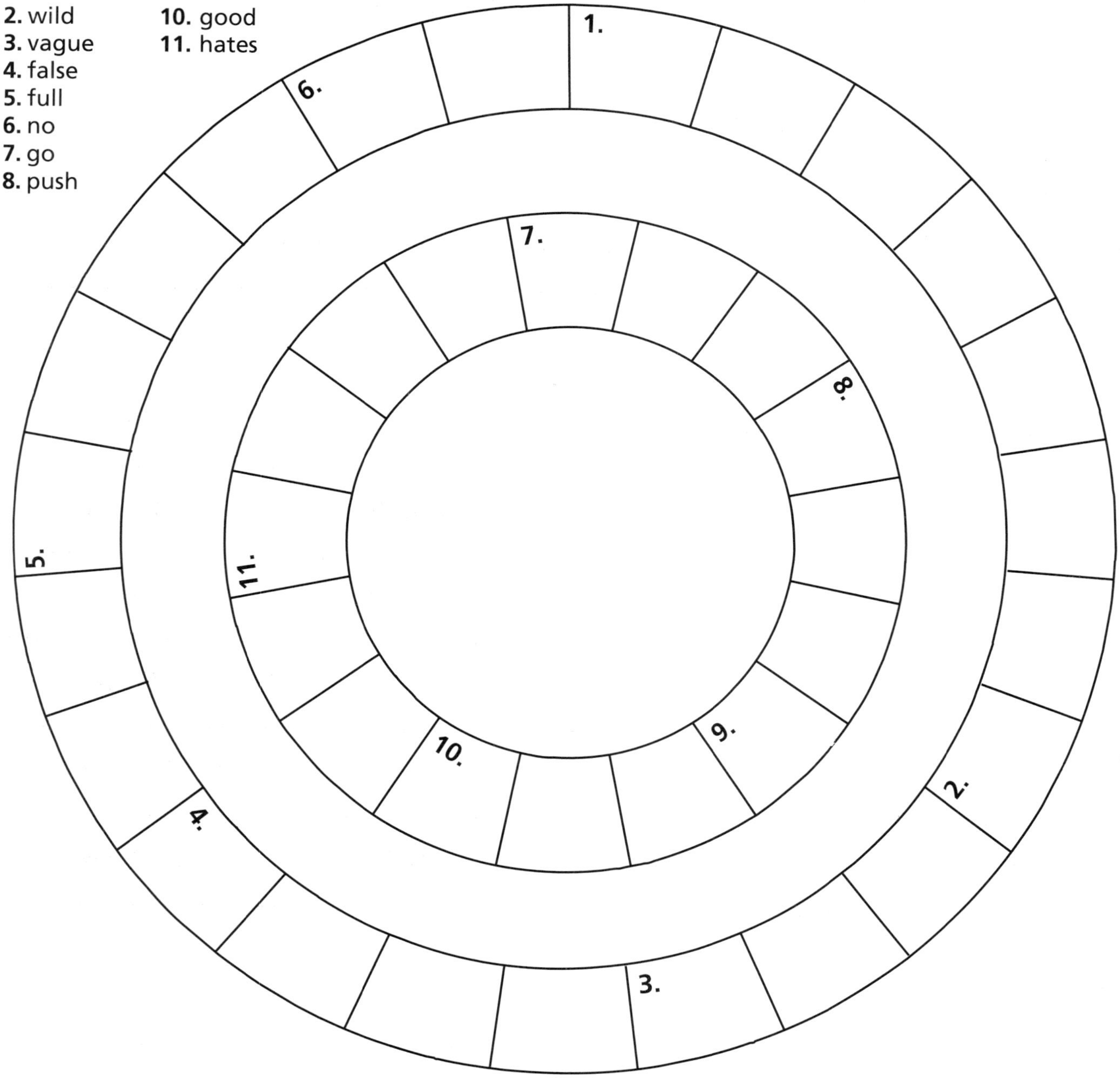

Just for Fun
Make up an antonym bull's-eye puzzle for a classmate to solve.

Name ______________________________ Date ______________

Riddle #1

PHONICS, CONTRACTIONS, RHYMING WORDS, PRONOUNS, PROPER NOUNS, AND SYLLABLES

Follow the directions below to find the answer to the riddle. Circle the words that have not been crossed out. Read the answer going down from top to bottom, starting in the upper left-hand corner of the grid. Write the answer to the riddle on the lines.

What did the beach say when the tide finally came in?

- Cross out all words that start and end with vowels.
- Cross out all words that are pronouns.
- Cross out all words that are contractions.
- Cross out all words that are proper nouns.
- Cross out all words that have exactly five syllables.
- Cross out all words that rhyme with *peach*.
- Cross out all words with identical double consonants in the middle.

LIABILITY	YOUR	TIME	BEACH
ARENA	HASN'T	TEXAS	NO
REACH	JASON	EXAMPLE	MIDDLE
WASN'T	LONG	AUDITORIUM	SEA
YOURSELF	AUTOMOBILE	LITTLE	THEIR
MYSELF	TEACH	VEGETARIAN	ENCYCLOPEDIA
SATURDAY	WOULDN'T	INVITE	BURROW

Answer to Riddle

____

____

Name ______________________________ Date ______________

Riddle #2

Spelling and Homophones

Follow the directions below to find the answer to the riddle. Circle the words that have not been crossed out. Read the answer going down from top to bottom, starting in the upper left-hand corner of the grid. Write the answer to the riddle on the lines.

When is a station wagon not a station wagon?

- Cross out all words that have an *ie* spelling.
- Cross out all words that are homophones for the word *vain*.
- Cross out all words that start and end with the same letter.
- Cross out all words that name things that are used to measure something.
- Cross out all words that have exactly three syllables.
- Cross out all words that begin with the word *up*.
- Cross out all words that end with the suffix *-sion*.

EXTENSION	GRIEF	TURNS	NIECE
UPHEAVAL	IT	VANE	A
GRAFTING	STATIONS	INTO	HAPPILY
BAROMETER	DISCUSSION	REPORTER	DRIVEWAY
FRIEND	DIVISION	THERMOMETER	UPHOLSTER
VEIN	NOTATION	RAPIDLY	EFFICIENT
WHEN	ODOMETER	UPSTREAM	RULER

Answer to Riddle

____

____

Name ____________________ Date ____________

Riddle #3

CONJUNCTIONS, ADVERBS, SYNONYMS, AND SYLLABLES

Follow the directions below to find the answer to the riddle. Circle the words that have not been crossed out. Read the answer going down from top to bottom, starting in the upper left-hand corner of the grid. Write the answer to the riddle on the lines.

Why did the boy bring a ladder to his middle school?

- Cross out all words that are conjunctions.
- Cross out all words that are adverbs.
- Cross out all words that are the names of colors.
- Cross out all words that are synonyms of the word *weak*.
- Cross out all words that have exactly four syllables.
- Cross out all words that are names of things to read.
- Cross out all words that begin and end with the letter *e*.
- Cross out all words that are names of habitats.

AZURE	GREGARIOUS	TO	HIGH
PANORAMA	MARSH	PRAIRIE	ALMANAC
BROCHURE	BUT	MAGAZINE	POPULATION
FEEBLE	FRAIL	BE	ELEVATE
DESERT	GLADLY	FRAGILE	SCHOOL
HE	WANTED	EXCESSIVE	MAUVE
CAREFULLY	OR	IN	AND

Answer to Riddle

____

____

Name ______________________________ Date ______________

Riddle #4

PHONICS, PAST PARTICIPLES, ANTONYMS, AND VOCABULARY

Follow the directions below to find the answer to the riddle. Circle the words that have not been crossed out. Read the answer going down from top to bottom, starting in the upper left-hand corner of the grid. Write the answer to the riddle on the lines.

Why don't pigs make good drivers?

- Cross off all words that start and end with the same vowel.
- Cross off all words that are past participles.
- Cross off all words that are the names of trees.
- Cross off all words that are proper nouns.
- Cross off all words that have exactly four syllables.
- Cross off all words that are antonyms of the word *happy*.
- Cross off all words that name objects that hold things.
- Cross off all words that are synonyms of the word *hot*.

TAKEN	SPRUCE	INTERESTING	AGENDA
ENGAGE	ENTERTAINMENT	ARE	HOGS
DEPRESSED	JENNIFER	WILLOW	SEEN
BECAUSE	BLISTERING	SORROWFUL	ARMADA
BOSTON	GLUM	VASE	SCORCHING
BIRCH	THEY	BROKEN	KENNEDY
UNDERSTANDING	CANISTER	ROAD	BEGUN

Answer to Riddle

____

____

Answer Key

Ferris Wheel Fun (page 5)

Answers will vary. Possible answers include:

eight	gin	high	hint	hip	hire
his	hit	ire	night	nip	nit
per	pie	pier	piers	pig	pigs
pin	pint	pit	rep	rig	right
rigs	ripe	sight	sin	sip	sir
sire	thin	this	tie	tier	tight
tin	tinier	tip	tire		

Circle Words (page 6)

Answers will vary. Possible answers include:

are	art	arts	aware	awe	ear
ewe	hear	heart	hearts	her	here
hew	hoe	hoes	how	ohs	owe
owes	raw	rear	reset	rest	rests
retest	retests	sea	sear	set	sets
sew	she	shear	shoe	shoes	show
shower	stew	straw	tea	tear	test
tester	tests	war	ware	wares	wart
wear	west	wet	wets	woe	woes

Pepperoni Pizza (page 7)

Answers will vary. Possible answers include:

are	art	arts	aye	era	ire
pep	per	persist	persists	pert	pray
prep	pry	rare	rarer	rat	rats
ray	rep	rip	ripe	riper	sip
sir	sirs	star	stars	stay	stray
stripe	tar	tars	tart	tarts	try
tray	trip	tripe			

Pick a Petal (page 8)

Answers will vary. Possible answers include:

ale	ales	ape	apes	bed	beds
belt	bet	deal	dealt	debt	did
die	dies	diet	idea	ides	lap
lapel	leap	led	let	pal	pale
pales	pea	peal	pelt	pep	pie
pies	sea	seal	set	tea	teal

More Pick a Petal (page 9)

Answers will vary. Possible answers include:

amp
amps
arm
asp
ate
ear
earn
eat
eaten
man
mane
map
maps
mar
mast
masts
mat
mate
mats
nap
naps
near
neat
neaten
net
nets
pan
par
past
paste
pat
pate
ram
ramp
ramps
ran
rap
raps
rasp
rasps
rat
rate
rats
sane
sap
sat
stamp
stamps
star
steam
tap
taps
tar
taste
tea
team
tear
ten

Word Quilt (page 10)

Answers will vary. Possible answers include:

ape
aped
apes
art
ate
away
awry
bear
bed
bet
beta
but
clink
clip
clips
dear
debt
ear
eat
gag
gap
gape
gapes
gaping
gate
gated
gates
gay
gin
ink
inking
kin
king
lie
lies
link
lip
lips
lisp
lisps
mink
oar
oat
par
pat
pate
pates
paw
pay
pea
pear
pep
pet
pie
pies
pin
ping
pinging
pink
pinking
quota
quote
rag
raging
rap
rat
raw
ray
rot
rote
set
sip
sips
sling
slink
tap
taping
taps
tar
tartar
tat
tea
tot
tub
tube
tubes
wag
waging
war
wart
way
wrap
wrote
yap

Magic Square #1 (page 11)

Answers will vary. Possible answers include:

ale
arc
are
art
arty
bore
breach
came
camel
car
cart
char
chart
chat
cub
curb
ear
eat
eater
enter
eye
gear
gel
gem
gent
get
glare
gleam
glen
hale
ham
hamlet
hare
hat
hate
ire
lame
late
leg
let

male
mare
mart
meat
men
mere
met
metro
near
net
oral
ore
pit
pity
ram
rare
rarer
rat
rate
rent
retro
rip
rob
robot
rot
rub
tale
talent
tar
tart
ten
tent
tip
tire
tore
tram
treat
year
yet

Magic Square #2 (page 12)

Answers will vary. Possible answers include:

ace
again
against
aging
angst
bean
bug
bugs
bun
buns
bus
bust
busts
caging
cam
can
cane
cans
dig
digs
din
ding
dining
dip
dog
dogs
doing
dot
dots
drip
engine
gain
gains
gay
gig
gigs
gin
gins
gnu
gnus
got
gust
gusts
mace
man
map
may
nag
nags
neap
nine
nip
nub
nun
nuns
pace
pain
pains
pan
pane
pang
pangs
pans
pay
paying
pig
pigs
pin
pine
ping
pings
pins
prince
rid
rig
rigs
ring
rings
riot
riots
rip
snag
snags
snip
sniping
snub
sun
suns
toga
tot
tots
yam
yang
yin

Magic Square #3 (page 13)

Answers will vary. Possible answers include:

ail
art
arts
avail
bah
bam
bar
bard
bards
bat
bath
bats
don
done
drab
drat
drown
ewe
eye
foe
for
ford
fort
forth
forts
fowl
frat
frats
halt
halts
ham
hard
hat
hats
hut
huts
knew
know
lam
lamb
lard
lards
lone
lord
low
lye
malt
malts
mar
mars
mat
math
mats
new
nod
nods
nor
north
now
odor
oral
owe
owl
rah
rail
ram
rat
rats
rod
rods
row
shut
star
stars
start
starts
stat
stats
tailor
tailors
tat

Magic Square #4 (page 14)

Answers will vary. Possible answers include:

aha
ail
aim
aiming
asp
asps
flit
flits
float
floats
gin
gist
has
hat
hating
hats
hot
limit
limiting
limits
line
liner
liners
lines
lit
lot
lots
mine
mines
mining
mist
misting
mists
nerve
nerves
nest
nesting
nests
never
nine
nines
nit
nits
oat
oats
pat
pats
resist
rest
resting
rests
sail
sap
saps
sat
satin
save
saver
savers
saves
silt
sin
sine
sing
singing
sins
sis
sit
sits
snit
snits
spat
spats
stain
staining
stains
stat
tail
tap
taps
tasting
that
thaw
tilt
tilting
tilts
timing
tin
tine
tines
tins
tot
tots
vat
vats

Valentine's Day (page 15)

Answers will vary. Possible answers include:

advent
advents
aid
aids
ail
ails
ale
ales
ant
ants
ate
daily
dainty
dale
dales
date
dates
day
days
deal
deals
den
dens
dent
dents
destiny
detain
detains
dial
dials
die
dies
dine
dines
dive
dives
ease
eased
eat
eaten
eats
edit
edits
eel
eels
end
ends
eve
eves
eye
eyes
invade
invades
invent
invents
lad
lads
laid
late
lava
lay
lays
lead
leads
lean
leanest
leans
leaves
lend
lends
lent
lid
lids
line
lined
lines
lint
nail
nails
nay
nays
neat
neatly
nest
net
nets
nine
nines
nit
nits
sad
said
sail
sailed
salad
save
saved
say
seal
sealed
send
sent
silent
slant
slave
slide
snail
snide
stain
stained
stand
stay
tad
tail
tails
tale
tales
tan
tans
tea
teas
tease
teased
ten
tend
tends
tens
tense
tin
tine
tines
tins
vain
vane
vanes

vat
vats
veil
veils
vein
veined
veins
vent
vents
vest
vet
vets
vial
vials
vine
vines
yea
yen
yes

Abraham Lincoln (page 16)

Answers will vary. Possible answers include:

acorn
aha
ail
arm
bail
ball
balm
bam
ban
banana
bill
bin
bra
brain
bran
call
calm
cam
can
coal
coin
corn
hail
hall
ham
harm
hill
hoar
horn
lamb
mall
man
manila
mill
moan
mob
nab
nail
nil
nor
oar
oral
orb
rah
rail
rain
ram
ran
rind
roam
rob

George Washington (page 17)

Answers will vary. Possible answers include:

age
ages
aging
are
ash
aster
ate
awe
eat
eaten
eating
eats
egg
eggs
eight
engage
engaging
ewe
ewes
gag
gage
gages
gags
gain
gains
gang
ganging
gangs
get
gets
gin
ginger
gins
gorge
gorges
gorging
got
grain
grains
groin
groins
hag
hags
hang
hanger
hangers
hangs
has
hew
hews
hinge
hinges
hinging
into
ire
nag
nags
near
nears
nor
north
oat
oats
ogre
ogres
one
ones
rag
rage
rages
raging
rags
rain
rains
ran
rang
range
ring
rings
row
rows
sag
sage
sang
sea
seat
see
seen
sew
sewer
sign
singe
sin
sing
singer
snag
soar
son
song
sore
stone
store
storing
stow
string
strong
tan
tans
than
then
throne
thrones
tin
tine
tines
ton
tone
tones
toning
tons
train
trains
twin
twine
twines
twins
wag
wage
wages
wagging
wags
wash
waste
wear
wears
win
wine
wines
wins
worn

Mother's Day (page 18)

Answers will vary. Possible answers include:

ads
are
arm
arms
army
aster
ate
aye
ayes
dam
dame
dames
dams
dare
dares
day
days
dear
dears
dot
dote
dotes
dots
dry
ear
ears
era
eras
had
ham
hams
hard
hardy
hay
head
heads
hear
hears
hem
hems
her
hey
home
homer
homers
homes
hot
mad
made
mare
mares
may
moat
moats
mode
modes
more
moth
moths
oat
oats
ore
ores
other
rat
rate
rates
rats
ray
rays
read
reads
ready
ream
reams
red
reds
rest
roast
rot
rote
rots
sad
say
seam
shade
sod
some
sore
soy
stay
steady
stem
store
story
stream
tad
tardy
the
thread
threads
toe
toes
toy
toys
try
yard
yards
yea
year
years
yes
yet

Father's Day (page 19)

Answers will vary. Possible answers include:

ads
afar
after
ahs
are
art
arts
aster
ate
aye
ayes
dare
dares
dash
data
day
days
deaf
dear
dears
draft
drafts
dry
ear
ears
era
eras
fad
fade
fades
fads
far
fare
fares
fast
faster
fat
fate
fates
father
fathers
fats
fear
fears
fed
fey
frat
frats
fray
frays
had
hardy
has
hate
hates
hay
head
heads
hear
hears
her
hey
raft
rafts
rat
rate
rated
rates
rats
ray
rays
read
reads
ready
red
reds
rest
sad
safe
safer
say
sear
shade
star
stare
stay
steady
tad
tardy
tear
tears
the
thread
threads
tread
treads
try
yard
yards
yea
year
years
yes
yet

Summer Vacation (page 20)

Answers will vary. Possible answers include:

ace
aces
action
arc
arcs
are
arm
arms
ate
cam
came
cams
cane
canes
cat
cats
cave
caves
come
comes
cut
cute
cuter
cuts
ear
ears
eat
eats
eave
eaves
ice
ices
ire
its
man
mar
mare
mares
mat
mate
mates
mats
meant
meat
meats
men
mine
mines
mum
near
nears
nor
norm
norms
ram
rams
ratio
ratios
rim
rims
rum
rums
run
runs
sac
same
sate
save
saver
sea
seam
seat
simmer
sin
sit
snore
sum
summer
tame
tamer
tames
tan
tans
tar
tars
tea
tear
tears
teas
time
timer
times
tin
tine
tines
vacant
vacation
van
vane
vanes
vans
vent
vents
vest
vice
vices
vine
vines

Independence Day (page 21)

Answers will vary. Possible answers include:

ace
aced
add
added
addend
aid
aide
aided
and
ape
aped
aye
cad
caddy
cay
dad
daddy
dance
danced
day
dead
deaden
dean
deed
deeded
deep
deepen
den
depend
depended
dice
diced
die
died
dine
dined
dip
end
ended
ice
iced
icy
idea
independence
inn
nap
nay
neap
nine
nip
pace
paced
pad
padded
paddy
paid
pain
pained
pan
panned
pay
pea
pen
penned
pie
pin
pine
pined
pinned
yap
yen

Happy Halloween (page 22)

Answers will vary. Possible answers include:

ale
all
allow
anew
ape
apple
apply
awe
eel
elope
ewe
eye
hale
hall
happen
happy
hay
heap
heel
hello
help
hen
hey
hoe
hole
holly
honey
how
howl
lane
lap
law
lawn
leap
loan
lone
lonely
lop
low
lowly
nap
nay
neap
new
nope
one
open
owe
owl
own
pal
pale
pall
pan
pane
panel
paw
pawn
pay
peal
peel
pen
pew
plan
plane
poll
pollen
pone
wall
wan
wane
way
weal
wean
wee
weep
well
whale
wheel
whelp
when
whole
wholly
yap
yaw
yawn
yea
yell
yellow
yelp
yen
yowl

Festival of Lights (page 23)

Answers will vary. Possible answers include:

aft
age
ages
ahs
ail
ails
ale
ales
alight
all
ate
east
eat
eats
eave
eaves
elf
evil
evils
fail
fails
fast
feast
feasts
fig
figs
file
files
fill
fills
filth
fish
fishes
fist
fists
fit
fits
five
flag
flags
flight
float
floats
fog
follies
gait
gal
gale
gales
gall
gals
gas
gate
gates
get
gets
ghost
ghosts
gift
gifts
gill
gills
gilt
glass
gloat
gloats
glove
gloves
goal
goals
goat
goats
golf
golfs
got
hag
hags
hail
hails
half
hall
halls
halt
halves
haste
hat
hate
hats
have
hello
hill
hills
hilt
hit
hive
hives
hog
hole
holes
hose
host
hostel
hostile
hosts
hot
hotel
hotels
hovel
ill
its
lag
lags
lass
last
least
leg
legs
list
lists
loaf
loft
log
logs
lost
lot
lots
oaf
oat
oats
off
oil
oils
oval
ovals
sag
sage
sages
sail
sat
sate

save
saves
sea
seal
seals
seas
self
set
sets
shaft
shafts
shale
shall
shot
shovel
sift
sifts
sight
silt
sis
sit
sits
slate
slates
slave
slaves
slight
slot
slots
stag
stage
stages
stags
stale
steal
stile
stiles
still
stove
stoves
tag
tags
tail
tails
tale
tales
tall
tallies
tea
teal
test
tests
tight
tights
tile
tiles
tilt
tilts
toast
toasts
toga
togas
toll
vale
vales
valet
valets
veal
veil
veils
vest
vests
vial
vials
vote
votes

Merry Christmas (page 24)

Answers will vary. Possible answers include:

aim
aims
air
airs
airy
are
arm
arms
art
arts
ash
ashes
aster
asters
ate
cam
came
cams
cat
cats
chair
chairs
char
chars
chat
chats
cherry
cream
creams
cries
crime
crimes
cry
ear
ears
era
eras
hair
hairs
ham
hams
hare
hares
harm
harms
has
hat
hate
hates
hats
hay
hear
hears
hem
hems
her
hers
him
ice
ices
icy
ire
mar
mars
mass
mast
masts
match
matches
may
meat
meats
merry
mist
mists
rah
raise
raises
rare
rarer
rate
rates
rat
rats
ray
rays
reach
ream
reams
rhyme
rhymes
rim
rims
same
say
shirt
shirts
sir
sire
sires
stair
stairs
star
stare
stars
stir
stream
streams
tea
teach
tear
tears
teas
they
time
times
tire
tires
tram
trams
trash
trashes
tries
try
yam
yams
year
years

Bone Search (page 25)

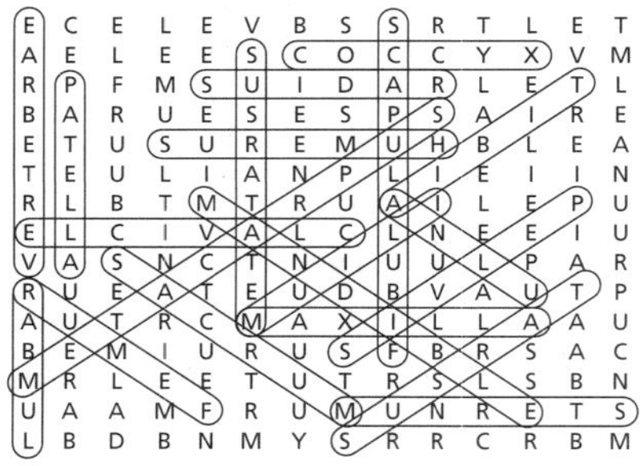

Machine Search (page 29)

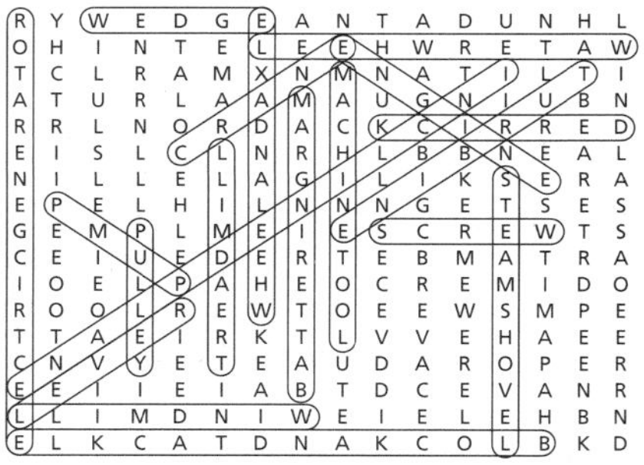

Capital Search (page 33)

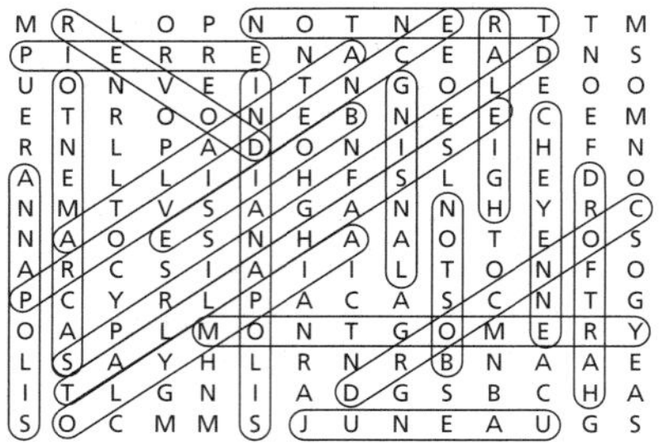

Color Search (page 26)

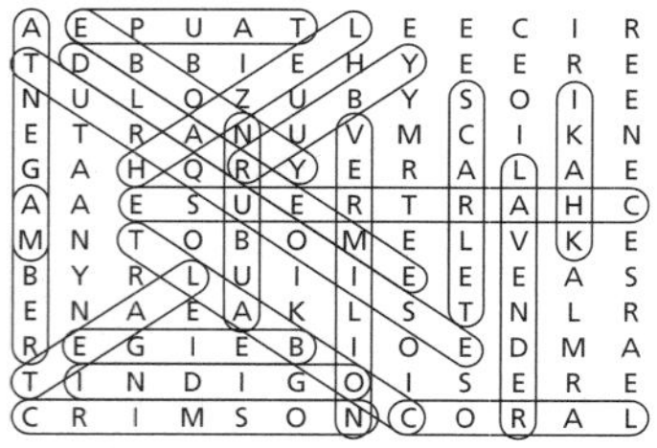

Mammal Search (page 30)

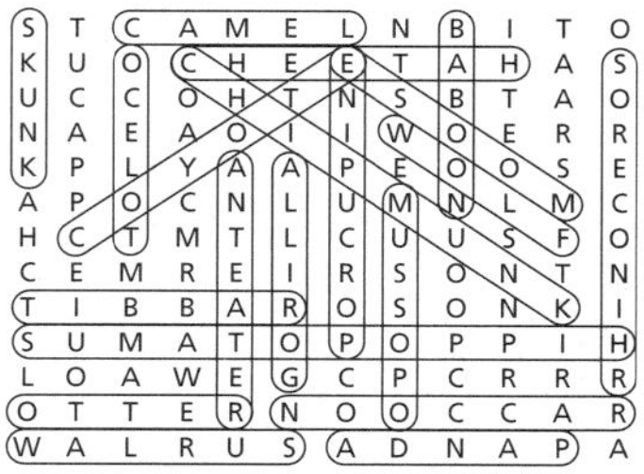

Colonial Search (page 34)

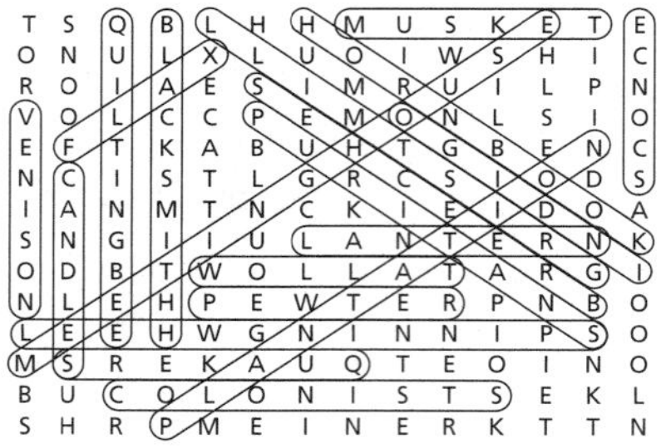

Dog Search (page 27)

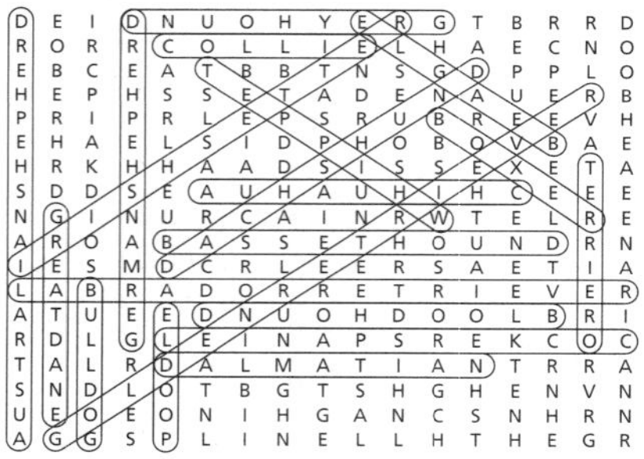

Space Search (page 31)

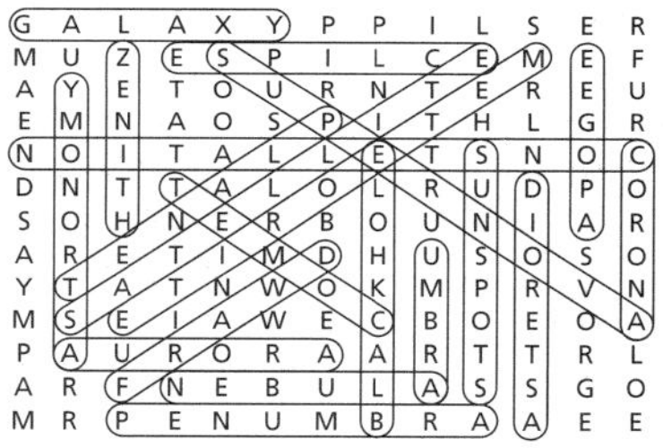

Country Search (page 35)

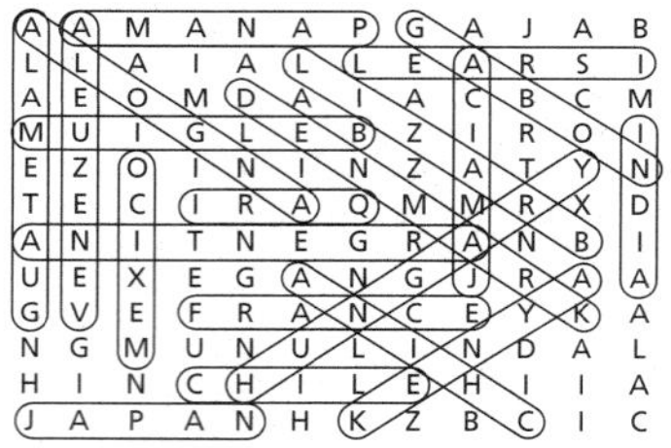

Flower Search (page 28)

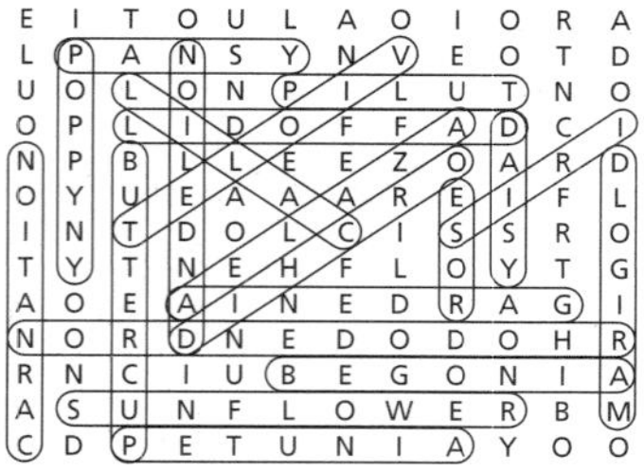

Weather Search (page 32)

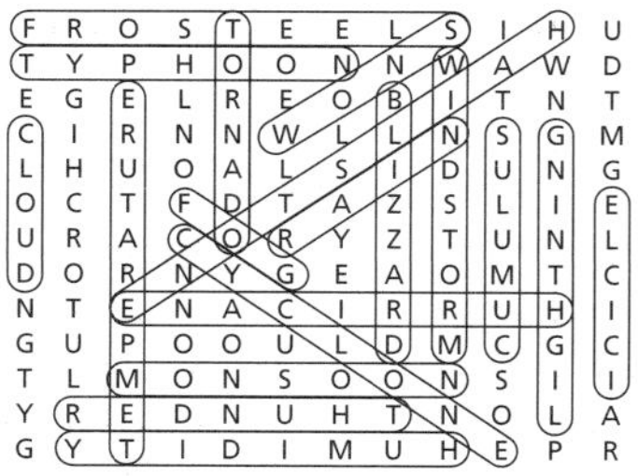

Geography Search (page 36)

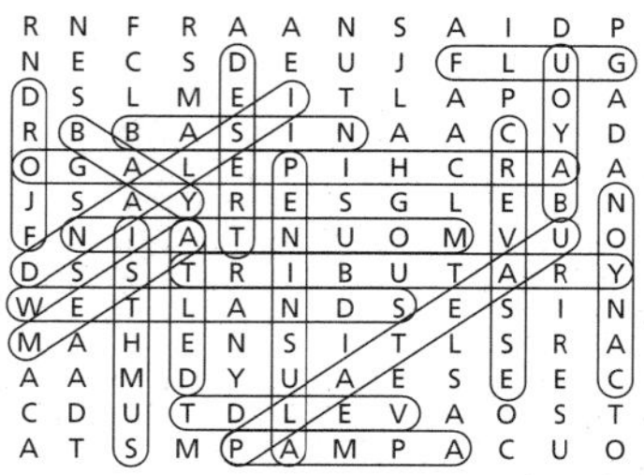

Native American Search (page 37)

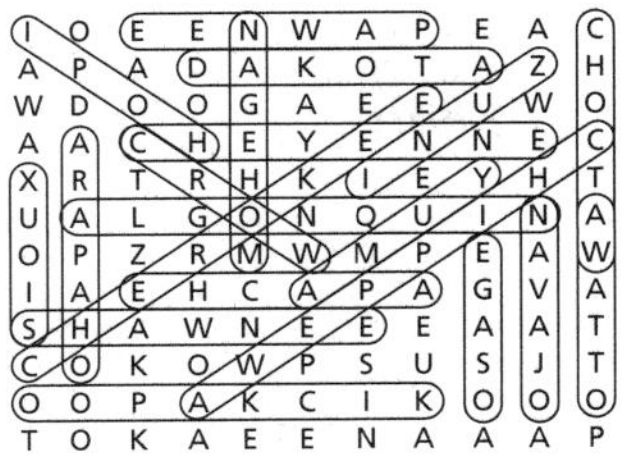

President Search (page 38)

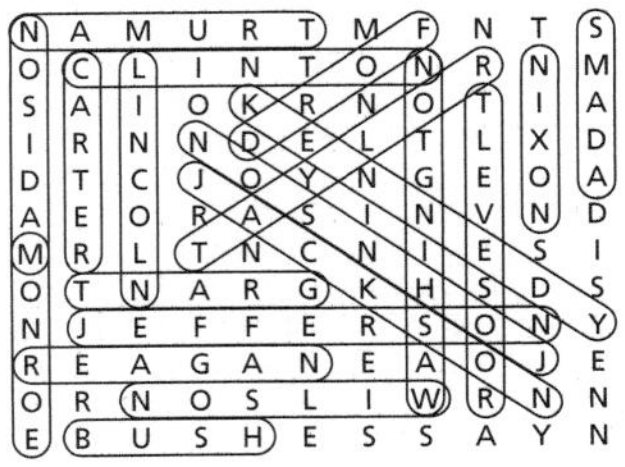

State Search (page 39)

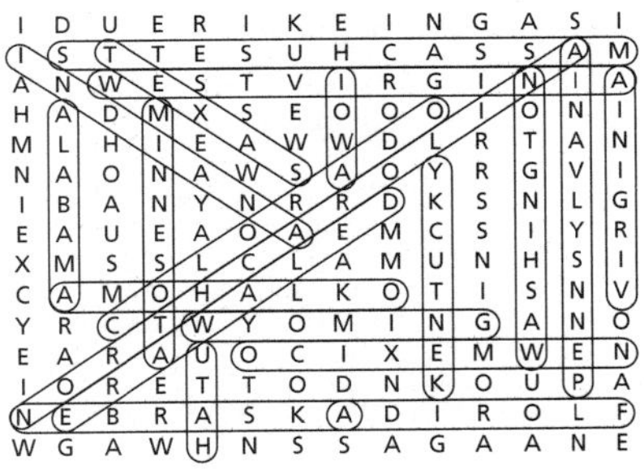

City Search (page 40)

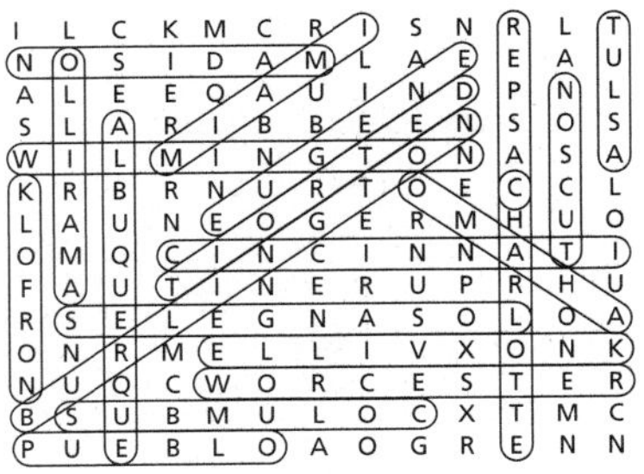

Category Games (pages 41–50)

Answers will vary.

Synonym Puzzle (page 51)

Across
2. quick
4. simple
6. tardy
8. disappear
10. near
12. foolish
14. repair

Down
1. quiver
3. tiny
5. answer
7. purchase
9. empty
11. odor
13. primates

Antonym Puzzle (page 52)

Across
2. worst
4. forbid
6. different
8. tame
10. narrow
12. night

Down
1. smooth
3. dirty
5. fail
7. never
9. below
11. many
13. large

Homophone Puzzle (page 53)

Across
2. capital
4. flower
6. bread
8. knead
10. allowed
12. hear
13. piece

Down
1. hanger
3. blew
5. aisle
8. knew
7. dough
9. write
11. higher

Past Participle Puzzle (page 54)

Across
2. risen
4. done
6. broken
8. begun
10. thrown
12. written
13. sunk

Down
1. stolen
3. chosen
5. forgotten
7. grown
9. swum
11. bitten

Noun Puzzle (page 55)

Across
2. speck
4. pioneer
6. knuckle
8. wish
10. path
12. orbit

Down
1. brink
3. progress
5. periscope
7. coward
9. neighbor
11. teen

Plural Puzzle (page 56)

Across
2. halves
4. children
6. pennies
8. berries
10. geese
12. potatoes

Down
1. matches
3. studios
5. chimneys
7. beliefs
9. mice
11. salmon
13. monkeys

Adjective Puzzle (page 57)

Across
4. cruel
6. glossy
8. various
10. afraid
12. grumpy

Down
1. firm
2. lazy
3. eager
5. circular
7. obvious
9. hectic
11. phony

Abbreviation Puzzle (page 58)

Across
2. foot
5. tablespoon
6. gallon
8. pint
10. inch
11. ounce
12. quart

Down
1. dozen
3. pound
4. mile
5. teaspoon
7. hour
9. year

Mammal Puzzle (page 59)

Across
2. leopard
4. porcupine
6. cheetah
8. tiger
9. bear
10. panda
11. gorilla

Down
1. giraffe
3. camel
5. kangaroo
7. elephant

Animal Offspring Puzzle (page 60)

Across
2. fawn
4. kid
6. colt
8. chick
10. lamb
11. gosling
12. joey

Down
1. eaglet
3. duckling
5. kitten
7. calf
8. cub
9. piglet

Plant Puzzle (page 61)

Across
2. pollen
4. evergreens
6. botanist
8. chlorophyll
10. stigma
12. cactus

Down
1. redwoods
3. petals
5. root
7. nectar
9. leaf
11. stem

Insect Puzzle (page 62)

Across
1. colony
3. feelers
5. cricket
7. larva
8. six
9. swarm
10. bee

Down
1. caterpillar
2. hive
3. flea
4. termite
6. two

Bird Puzzle (page 63)

Across	Down
2. nest	1. rooster
4. pink	3. beak
6. feathers	4. peacock
7. parrot	5. red
8. bald eagle	7. penguin
9. bird	

Number, Please Puzzle (page 64)

Across	Down
2. sixty	1. ten
4. sixteen	3. three
6. twelve	5. seven
8. four	6. three
10. twelve	7. two
	9. twenty

What Would You Do with a Ballad? (page 65)

1. b	**6.** a
2. d	**7.** d
3. d	**8.** b
4. b	**9.** c
5. c	**10.** a

What Would You Do with a Cashew? (page 66)

1. c	**6.** a
2. a	**7.** d
3. b	**8.** c
4. d	**9.** a
5. b	**10.** d

What Would You Do with Galoshes? (page 67)

1. d	**6.** c
2. b	**7.** b
3. b	**8.** d
4. d	**9.** a
5. a	**10.** c

What Would You Do with a Javelin? (page 68)

1. d	**6.** b
2. c	**7.** d
3. b	**8.** a
4. d	**9.** c
5. d	**10.** a

What Would You Do with a Kaleidoscope? (page 69)

1. c	**6.** b
2. a	**7.** c
3. d	**8.** a
4. a	**9.** d
5. b	**10.** c

What Would You Do with a Legume? (page 70)

1. b	**6.** a
2. b	**7.** b
3. a	**8.** a
4. d	**9.** c
5. c	**10.** d

What Would You Do with a Niche? (page 71)

1. d	**6.** a
2. b	**7.** c
3. c	**8.** d
4. a	**9.** a
5. b	**10.** c

What Would You Do with a Route? (page 72)

1. b	**6.** b
2. a	**7.** a
3. d	**8.** b
4. c	**9.** c
5. b	**10.** a

What Would You Do with a Toucan? (page 73)

1. d	**6.** b
2. c	**7.** b
3. b	**8.** d
4. d	**9.** c
5. a	**10.** a

What Would You Do with a Vandal? (page 74)

1. b	**6.** d
2. c	**7.** a
3. a	**8.** a
4. d	**9.** d
5. c	**10.** b

Ladder Links #1 (page 75)

1.	**2.**
hole	bears
hold	fears
hood	feats
hoop	flats
coop	flags

Ladder Links #2 (page 76)

1.	**2.**
five	foot
file	food
fill	fond
fall	bond
fail	band
tail	hand

Ladder Links #3 (page 77)

1.	**2.**
fire	stile
hire	stale
here	stare
herd	stars
head	sears
heat	tears

Ladder Links #4 (page 78)

1.	**2.**
smart	hook
start	look
stark	loot
stack	lost
slack	list
black	fist
	fish

Ladder Links #5 (page 79)

1.	**2.**
hope	slow
hole	slot
hold	slat
gold	seat
goad	sent
grad	pent

Ladder Links #6 (page 80)

1.	**2.**
never	hard
lever	ward
liver	wand
lives	wane
limes	cane
times	case
	ease

Ladder Links #7 (page 81)

1.	**2.**
chew	blank
chow	bland
show	brand
slow	braid
slew	brain
sled	grain

Ladder Links #8 (page 82)

1.	**2.**
beat	warn
meat	warm
melt	ward
malt	card
male	cord
mane	cold

Ladder Links #9 (page 83)

1.	2.
tool	sage
toil	page
foil	pale
fail	pall
fall	wall
ball	well
bale	

Ladder Links #10 (page 84)

1.	2.
coal	whale
foal	shale
fool	share
food	shore
fond	spore
find	sport
fine	spout

Noun Analogies (page 85)

1. b
2. d
3. c
4. a
5. b
6. c
7. d
8. c
9. b
10. d

Verb Analogies (page 86)

1. b
2. d
3. c
4. d
5. a
6. d
7. b
8. c
9. c
10. d

Adjective Analogies (page 87)

1. d
2. a
3. b
4. d
5. b
6. d
7. b
8. a
9. b
10. d

Analogy Challenge #1 (page 88)

1. b
2. a
3. c
4. d
5. b
6. d
7. c
8. d
9. b
10. a

Analogy Challenge #2 (page 89)

1. d
2. c
3. d
4. a
5. b
6. c
7. d
8. b
9. a
10. c

Analogy Challenge #3 (page 90)

1. a
2. d
3. b
4. c
5. b
6. c
7. b
8. c
9. a
10. b

Analogy Challenge #4 (page 91)

1. c
2. d
3. b
4. d
5. c
6. d
7. a
8. b
9. c
10. a

Synonym and Antonym Analogies (page 92)

1. d
2. a
3. b
4. c
5. b
6. b
7. a
8. c
9. a
10. d

On the Menu Analogies (page 93)

1. a
2. b
3. d
4. c
5. b
6. d
7. a
8. c
9. a
10. b

Sports Analogies (page 94)

1. d
2. b
3. d
4. c
5. d
6. a
7. b
8. c
9. d
10. a

Adventure Analogies (page 95)

1. b
2. a
3. d
4. b
5. c
6. b
7. a
8. c
9. d
10. c

People Analogies (page 96)

1. b
2. a
3. c
4. a
5. b
6. c
7. b
8. d
9. b
10. d

Palindrome Fun (page 97)

1. dad
2. gag
3. radar
4. race car
5. eve
6. level
7. ewe
8. peep
9. refer
10. civic
11. mom
12. noon
13. toot
14. rotor
15. solos

Homophones (page 98)

1. J sum
2. F see
3. I sun
4. K bear
5. E deer
6. G fare
7. B pail
8. L cent
9. C nose
10. A hole
11. D stair
12. H tail

Hidden Animals (page 99)

1. horse
2. dog
3. bat
4. bear
5. mouse
6. gerbil
7. leopard
8. ewe
9. hen
10. goat
11. toad
12. ostrich

Hidden Foods (page 100)

1. pear
2. mango
3. turnip
4. tofu
5. tomato
6. peach
7. fish
8. bean
9. beef
10. lime
11. apple
12. toast

Alphabetical Bubbles (page 101)

Answers will vary.

Hidden Trios (page 102)

Answers will vary. Possible answers include:
are—stare, arena, pare, parent
hip—ship, hips, friendship
sit—sitter, situation, transit
top—topple, stop, atop, topiary
way—away, wayward, sway
ten—often, tennis, tender, attention

More Hidden Trios (page 103)

Answers will vary. Possible answers include:
lot—plot, allot, pilot
sea—seasick, season, sealed
own—gown, town, renown, owner
row—arrow, rower, rowboat, borrow, tomorrow
let—lettuce, outlet, letter
set—settlers, upset, setup

Four-Letter Riddle Words (page 104)

1. wind
2. hard
3. bent
4. join
5. hike
6. quit
7. fail
8. safe
9. brag
10. idea
11. rare
12. menu
13. mean
14. risk
15. tell
16. chin
17. chew
18. make

Answer to the riddle: When it is a drumstick.

Odd Word Out (page 105)

1. paint—not a color
2. encyclopedia—not part of a book
3. chef—not a method of preparing food
4. huddle—not a baseball term
5. sycamore—not a type of flower
6. curse—not a computer term
7. dollar—not a metal
8. foot—not part of the face
9. alligator—not a type of snake
10. goalie—not a basketball term

More Odd Word Out (page 106)

1. many—not a preposition
2. Ohio—not a U.S. city
3. atlas—not an interjection
4. tango—not a stringed instrument
5. Australia—not a country in Europe
6. yard—not a measurement of time
7. tripod—not a geometric shape
8. London—not the name of a continent
9. route—not a part of a plant
10. motorcycle—not a weather word

Wacky Riddles (page 107)

1. a sponge
2. a deck of cards
3. a towel
4. your name
5. a hole
6. your word
7. footsteps
8. the moon

Verb Tense Bull's-Eye (page 108)

1. eat
2. throw
3. wear
4. ring
5. grow
6. write

Synonym Bull's-Eye (page 109)

1. damp
2. promise
3. entire
4. excuse
5. end
6. decide
7. earn
8. noticed

Antonym Bull's-Eye (page 110)

1. straight
2. tame
3. exact
4. true
5. empty
6. yes
7. stop
8. pull
9. lose
10. evil
11. loves

Riddle #1 (page 111)

Long time no sea

Riddle #2 (page 112)

When it turns into a driveway

Riddle #3 (page 113)

He wanted to be in high school.

Riddle #4 (page 114)

Because they are road hogs